SCIENCE FUN IN CHICAGOLAND

SCIENCE FUN IN CHICAGOLAND

A Guide for Parents and Teachers

With Over 800 Resource Descriptions

THOMAS W. SILLS, PH.D.

Dearborn Resources
P. O. Box 59677
Chicago, IL 60659

**For the 176 hard working teachers
in my science teaching methods courses.**

SCIENCE FUN IN CHICAGOLAND. Copyright © 1995 by Thomas W. Sills. All rights reserved. Printed in the United States of America. No part of this book may be used or reproduced in any manner whatsoever including information storage and retrieval systems without written permission in writing from the publisher except in the case of brief quotations embodied in critical articles and reviews. For information contact the publisher.

Published by Dearborn Resources, P. O. Box 59677, Chicago, IL 60659.

The publisher and author of this book assume no legal responsibility for the completeness or accuracy of the contents of this book nor any legal responsibility for any increase or decrease in the value of any enterprises, whether for profit or non-profit, by reason of their inclusion or exclusion from this work. Contents are based on information believed to be accurate at the time of publication.

Neither Dearborn Resources nor the author have accepted payment from any firms or organizations for inclusion in this book.

FIRST EDITION

Cover design by Dickinson Associates

Publisher's Cataloging in Publication
(Prepared by Quality Books Inc.)

Sills, Thomas W.
    Science fun in Chicagoland : a guide for parents and teachers with over 800 resource descriptions / Thomas W. Sills. -- 1st ed.

    p. cm.
    Includes bibliographical references and index.
    ISBN 0-9644096-0-7

    1. Scientific recreations. 2. Science--Study and teaching--Illinois.   I. Title.

Q164.S55 1995   793.8
                            QBI94-21308

Table of Contents

Acknowledgements **ix**

Foreword by Bernie Bradley **xi**

How to Use This Book **xvi**

Chapter 01 **BOOKS** 1

    Resource Reference Books 1-7
    Books on Activities and Methods of Teaching 8-14
    Professional Books on Teaching Philosophy,
        Science History, and Science Reference 14-18
    Local Bookstores 19-23
    Mail Order Books For Sale 23-26

Chapter 02 **COMPUTERS** 27

    Computer Information Sources 27-30
    The Internet 31-32
    Computers and Education 32-35
    Local Computer and Software Suppliers 35-37
    Mail Order Suppliers 37-39

Chapter 03 **EDUCATION** 40

    Science Education Sources 40-41
    Schools and Academies K-12 42-44
    Science Education Opportunities 44-54

Chapter 04   EVENTS 55

      Local Events, Competitions, and Awards 55-60
      National Events, Competitions, and Awards 60-62

Chapter 05   EXCURSIONS 63

      Excursion Reference Books 63-64
      Local Excursion Opportunities 65-76

Chapter 06   GROUPS 77

      Science Groups Reference Book 77
      Local Science Groups 78-83
      National Science Groups 84-87

Chapter 07   INSTRUMENTS 88

      Scientific Measurement Instruments 88-91

Chapter 08   LIBRARIES 92

      Library Directories 92
      Local Libraries with Science Collections 93-102

Chapter 09   MATERIALS 103

      Hands-On Materials Reference Books 103-104
      Local Stores Selling Hands-On Materials and
         Science Equipment 104-110
      Mail Order Hands-On Materials and
         Science Equipment Sources 110-121

Chapter 10   PERIODICALS 122

      Periodical Directories 122-123
      Periodicals for Parents and Children 123-126
      Periodicals for Teachers 126-133
      Science Periodicals 133-140

Chapter 11 **SAFETY** 141

        Science Education Safety 141-144
        Mail Order Safety Equipment Suppliers 144

Chapter 12 **SCIENCE FAIRS** 145

        Science Fair Project Ideas 145-147
        Steps to a Winning Project Experiment 146
        Teacher's Guide to Local Science Fairs 148-149

Chapter 13 **TOYS** 150

        Science Toy Reference Books 150-151
        Educational Toy Loan Centers 151-152
        Local Toy Stores Selling Science Toys 152-155
        Science Toy Sources 155-165

Chapter 14 **VIDEO** 166

        Film and Video Sources 166-169
        Science Programs on Television 169-172

Index 175

About the Author 195

ACKNOWLEDGEMENTS

A countless number of people helped with the production of this book. They gave suggestions, answered telephone inquires, and made this book possible. It was overwhelming how individuals were willing to help and it would be impossible to acknowledge everyone here who openly and freely contributed to the information contained in this book.

A few individuals were more than generous with their time and energy to help the author. It is appropriate to list their names: Ethan Allen, Sherry Bushré, Maria Morales, Teachers Academy for Mathematics and Science; Christian Albert, Commonwealth Edison Power House; Steve Anzaldi, Kohl Children's Museum; Roger Bader, Von Steuben Metropolitan Science Center; Carolyn Blackmon, Department of Education, Field Museum of Natural History; Sr. Francis Berger, Loyola University Chicago Science Library; Adrian Beverly, Assistant Superintendent, Chicago Public Schools; Peter Blain and Michael Hyatt, Science Linkages in the Community; Judy Chiss and Diane Dickerson, Chicago Children's Museum; Chuck Clement, McCall Elementary School; Christine Crupa, Tilden High School; Susan Dahl, Education Office, Fermi National Accelerator Laboratory; Jennifer Fermon, SciTech; Gerald Foster, DePaul University; Ronald Gerut, Educational Technology Consultant; Howard Goldberg, University of Illinois at Chicago; Eric Hamilton, ACCESS 2000 and Chicago Systemic Initative; Lou Harnisch, Argonne National Laboratory; Tom Jensen, Bloom High School; Mary Keegan, Diane Schiller, Brian Wunar, Loyola University at Chicago; Zafra Lerman, Columbia College Chicago; Mrs. Macke, Morrill Elementary School; Lee Marek, Naperville North High School; Lane Phalen, Brigadoon Bay Books; Guen Pollack, Illinois State Board of Education; Anne Grall Reichel, Chicago Botanic Garden; Ken Rose, The Chicago Academy of Sciences; Jane Rosen, Golden Apple Foundation; Marilyn Russell, Brookfield Zoo and Chicago Zoological Society; Telkia Rutherford, Chicago Systemic Initiative, Chicago Public Schools; Mary Salzer, Piccolo Elementary School; Tom Senior, New Trier High School; Christine Smith, Hubbard High School; Nathan Unterman, Glenbrook North High School; Linda Wilson, Department of Education, John G. Shedd Aquarium; Melanie Wojtulewicz, Department of Instruction, Chicago Public Schools.

Thanks! Your efforts hopefully will be a great help to readers.

FOREWORD

Science is fun. Science is important. Science is essential to the survival of our society as we know it. Since science is essential to our survival, calls to improve the scientific literacy of our citizens have echoed throughout our country for the past four decades.

Friday, October 4, 1957, signaled the dawn of the space age with the launching of Sputnik by the U.S.S.R. While the birth of the United States science reform movement would come a few years later, that initiative was clearly conceived on that remarkable autumn day.

The specter of a Russian satellite orbiting the Earth was viewed in the U. S. as a national crisis. Concerns regarding American national security led to questions about national preparedness, the scientific literacy of its citizens, and the ability to compete on a global basis. These problems should have been identified in their early stages. Our society, however, too often operates in a reactive, crisis-oriented mode. Problems must reach emergency status before they are addressed. Three decades later, the focus has changed from national defense to the economic arena, but the questions remain the same.

One of the responses to this national crisis was a substantial increase in funding for the National Science Foundation for the purposes of expanding scientific research and improving science education. Part of the problem with producing scientifically literate students was that they were being taught by teachers who lacked confidence in their ability to teach science. Consequently, science was given little emphasis in the curriculum. In response, the early sixties saw the publication of what have come to be known as the "Alphabet" science programs. Programs, including Science Curriculum Improvement Study (SCIS), Elementary Science Study (ESS), and Science: A Process Approach (SAPA), focused on what science should be taught while emphasizing a "hands-on" approach to elementary science education. While the research indicated that the techniques employed by these programs were effective in improving achievement in science, they were not widely adopted. In most schools, the textbook was the curriculum and the most common instructional method was still the lecture.

As the American space program evolved during this decade, the crisis status of science education began to fade. The United States took the lead in space exploration and the concerns about national security and competitiveness waned. With no crisis, the reform of science education moved out of the spotlight and funding of curriculum development programs decreased. During the seventies there were warning signs to indicate that the efforts to improve science education in the schools were failing. *An Assessment of Science* conducted by the National Assessment of Educational Programs (NAEP) in 1978 reported that between the third and seventh grades, student interest in science decreased. Between the seventh and eleventh grades, student interest in science revealed an even greater drop off. Since reports such as these were not as dramatic as an orbiting Sputnik, science education did not receive great attention.

It would not be until 1983 that the reform of science education would again become a national priority. *A Nation at Risk* was the first of a series of reports documenting the problems of the American educational system and providing recommendations for its reform. It was consistently reported that American students were outperformed in science achievement by students in many other countries. The curriculum reform efforts of the sixties had not made much impact on the delivery of science education. Most teachers still lacked confidence in their ability to teach science and they relied on the textbook as their science curriculum. In this environment their students lost interest in science as a subject and as a career option. Thus, another crisis was born. Because so many students were abandoning their interest in science, the United States was facing a shortage in the fields of science and engineering.

At the same time, a number of changes had occurred on the American scene that would adjust the focus of future science reform initiatives. During the Sputnik era, scientific literacy meant producing a sufficient number of scientists and engineers to keep the United States competitive in the technological fields. As our society has faced an ever increasing number of technological related societal problems, the definition of scientific literacy has been broadened. Today, scientific literacy means science for all citizens. Genetic engineering, ozone depletion, nuclear power generation, greenhouse warming, acid rain, energy conservation, and other societal issues all reflect a science and technology component. If rational public policy decisions are to be made, scientific literacy must be a characteristic of the general population.

At the same time that the societal need for scientific literate citizens has increased, the United States' ability to compete in the global economic environment has been called into question. The work place has become more and more technology driven. A policy study commissioned by the Illinois State Board of Education predicts, "by the year 2010, virtually every job in the country will require some skill with information-processing technology."

Thus in light of past reform failures and research findings of cognitive psychology, a new wave of science education reform movements were initiated in the context of a changing economic environment.

Over the past ten years, three major science reform efforts have been started. Unlike the reform efforts begun in the post-Sputnik era, these reforms are not curriculum development programs. The American Association for the Advancement of Science developed *Project 2061*. The National Science Teachers Association developed *Scope, Sequence, and Coordination of Secondary School Science*. And, the National Research Council developed the *National Education Standards*. All three are curriculum design resources. Each program addresses all areas of a science curriculum. While past science reform curricula had addressed primarily what science content should be taught, these new initiatives enlarged their focus to include how science should be taught and how it should be assessed.

The recommendations made by each of these initiatives are research based and developed as our manufacturing based economy has given way to an information based economy. While each of these efforts takes a slightly different path, there are a number of consensus directions for science education programs:

- *Because the body of knowledge has grown at an exponential rate, it is not possible to teach students everything there is to know about science during their educational experience. The recommendation is to study a smaller number of ideas in much greater depth.*

- *Cognitive psychology reports that children are not blank slates when they enter school. Long before they pass through a school door for the first time, they have encountered many phenomena in the natural world. Children develop their own theories to explain the natural world. While some of these theories are naive, they are natural to the child. Just as children construct their knowledge before they start school, they continue to do so after they begin their formal education. Thus, a teacher's role is not to transfer facts and information from his or her brain to the student's. A teacher's role is to facilitate learning environments and experiences that will assist the student in the construction of accurate representations and understandings of the natural world.*

- *Learning theory relates that people learn most effectively when new concepts are related to previously learned ideas. If a student has no frame of reference for newly encountered material, then the material is not effectively learned. Common themes such as systems, patterns, structures and diversity are ideas that repeat themselves in all science disciplines. This has resulted in recommendations to teach science in the context of themes or big ideas.*

- *When children do science as scientists do, they can learn science and maintain an interest in science. Scientists design experiments to answer real questions. If the answers to questions are already known, there is no need for a scientist to conduct an experiment to get that answer. Unfortunately, students are far too often asked to conduct experiments for which there is only one right answer which is usually already known.*

- *Science should not just be an academic subject that students encounter for one period a day in school. Students live a natural and technological world that abounds with examples of scientific principles. Efforts must be made to relate the science that students practice in the classroom to the world that students encounter when they go home. The inertia of a bus ride, the chemistry of cooking, the physics of television, and the biology of sleep are just a few of the examples of science and technology that students find in their everyday world. Students should be encouraged to develop the ability to transfer the knowledge they construct in the classroom to the real world.*

These efforts are underway to help schools develop programs to improve the scientific literacy of students in this country. Is it purely the school's responsibility to achieve this important goal? Scientific research indicates that this responsibility must be shared.

Recently published research on the cognitive development of infants indicates that the first years of life are extremely important in providing a child with the maximum learning potential. Infants spend the first year of their life creating the nerve connections in their brains that will greatly affect their capacity to learn. From years one to twelve, children must continually exercise those nerve connections to prevent them from becoming inactive. The key to developing and maintaining nerve connections is stimulation. A baby must be raised in a stimulating environment. As a child grows and develops, the child must be exposed to and participate in a wide range of stimulating experiences. These experiences are the responsibility of the parents.

Strong partnership of the home and the school will produce our future scientifically literate citizens. Parents and teachers must do all that is possible to achieve this important goal. Children are naturally curious about their world. Most parents do maintain this curiosity so that students who are entering formal schooling still display it. The hardest job parents and teachers have is to nurture children's natural curiosity so that children will continue wanting to learn about their world.

How might one go about nurturing natural curiosity? What kind of experiences foster the development of scientifically literate citizens? Toys are the beginning of sustained scientific literacy. Manipulative toys are a fascination for children and for science teachers. The best science teachers have never lost that natural curiosity. These toys are also fascinating to scientists because the toys invite the question, "How does that work?" This question is basic to science whether it is being asked by a fourth grader, a fourth grade teacher, or a scientist at Fermi Lab. The need to be able to explain how the world works is the basis for the natural curiosity that defines science.

In the development of scientific literacy, books are a key component. Books can build imagination as words generate visual images. Books can provide information and inspire interest in careers.

Natural curiosity spontaneously develops in museums, nature centers, arboretums, conservatories, planetariums, and aquariums. These locations should all be required stops on a child's road to adulthood. Museums can create a sense of wonder in past accomplishments and expose young people to new adventures.

Does *Science Fun in Chicagoland* provide all the answers for parents and teachers who are interested in encouraging scientific literacy? No, that is not the purpose of the book. Since it contains an enormous data base of information about science, *Science Fun in Chicagoland* is a wonderful resource for parents and teachers. The book is a tool that can be used to locate stimulating materials and opportunities for the benefit of young people. How effective a tool this book will be will depend on the reader's initiative. Tom Sills has done the legwork needed to put this data base together. He is to be commended for the exhaustive research that was required to assemble this comprehensive information.

As the year 2000 approaches, the goal of the United States students ranking first in science achievement has been established as part of the *Goals 2000: Educate America Act*. If that goal is to be achieved, parents and teachers must each play a leadership role in the reform of science education. Parents must prepare their children for success in school and teachers must prepare themselves to provide their students with the best possible set of educational opportunities. *Science Fun in Chicagoland* is a valuable asset in that quest.

Bernie Bradley

Bernie Bradley is President of the Illinois Science Teachers Association, Science Specialist at Newberry Mathematics and Science Academy, Member of the Governor's Science Advisory Committee, Member of the Board of Trustees of the Teachers Academy for Mathematics and Science, 1991 recipient of the Illinois Distinguished Educator Award, Graduate of the Honors Science Teachers Program at Illinois State University, and 1988 recipient of the Golden Apple Award.

How to Use this Book

This book is your catalog to science in Chicago and nationwide.

Use the index to find addresses and telephone numbers.

Read each chapter to learn what is available.

Use as your shopping guide for gift and holiday buying.

Give as the perfect gift for that child in your family.

Plan a fun, adventuresome excursion for the weekend.

Find information quickly via books, computers and libraries.

Students, use this book to create a winning science project.

Teachers, use this resource as a field trip idea book.

Use as a shopping guide for school science equipment.

Develop science programs and lessons.

Find guest speakers on science subjects.

Give this book to your favorite teacher as an ideal gift.

"This book is priceless for teachers!" -- a teacher reviewer

SCIENCE FUN IN CHICAGOLAND

Chapter 01

Books

Resource Reference Books

These helpful resource lists and bibliographies can be found in libraries or purchased where indicated.

1993 CONSERVATION EDUCATION CATALOG
by the Conservation Education Advisory Board   Contact Kathy Engelson, Illinois Department of Energy and Natural Resources, 325 W Adams, Springfield, IL 62704   79 pages   Single copies may be obtained from the ENR Clearinghouse   800-252-8955
This catalog was previously published in 1990 as *Conservation Education in Illinois: A Sourcebook for Educators.* This extensive catalog lists resources under the topics of Agriculture & Land Conservation, Atmosphere, Ecology & Interdisciplinary, Energy, Fauna, Flora, Geology & Geography, Outdoor Recreation & Safety, Waste, Recycling & Polution Control, and Water. Over 80 resource publications and many agency addresses are also listed.

2 Science Fun in Chicagoland

BEST BOOKS FOR CHILDREN 1988-1991
edited by Sosa and Malcom   American Association for the Advancement of Science, 1333 H Street, NW, Washington, DC 20005   800-222-7809   1992   330 pages $ 38.95
This bibliographic resource book lists and describes over 800 science books for children categorized under seventeen different science subject areas.

BEST SCIENCE AND TECHNOLOGY REFERENCE BOOKS FOR YOUNG PEOPLE
by H. Robert Malinowsky   Oryx Press, 4041 N Central, Suite 700, Phoenix, AZ 85012-3397   800-279-6799   1991   216 pages $ 24.95
Written by a bibliographer of science and technology, this book describes 669 reference books by science subject with title, name, subject, and grade level indices at the end of the book. It also lists and describes general science bibliographies, biographical sources, dictionaries, encyclopedias, guides, histories, indexes, and yearbooks.

BEST SCIENCE BOOKS & AV MATERIALS FOR CHILDREN
by O'Connell, Montenegro, and Wolff   American Association for the Advancement of Science, 1333 H Street, NW, Washington, DC 20005   800-222-7809   1988
$ 20.00
This bibliographic reference work lists resource books under sixteen different science subject areas.

DIRECTORY OF STUDENT SCIENCE TRAINING PROGRAMS
Science Service, Inc., 1719 N St, NW, Washington, DC 20036   202-785-2255
Published annually. $ 3.00 per copy.
This directory lists extra-curricular opportunities in science for high ability precollege students.

DISCOVER OUR WORLD RESOURCE BOOK
Field Museum of Natural History Education Department   1985   Out of Print
This resource directory was developed through a collaborative effort of education leaders to describe Chicago's cultural institutions for educators. Check your local school library for a copy.

EDUCATORS GUIDE TO FREE SCIENCE MATERIALS - 35TH EDITION
edited by Mary H. Saterstrom   Educators Progress Service, Inc., 214 Center St, Randolph, WI 53956-1497   414-326-3126   1994   296 pages $ 27.95
This book lists and describes free science materials, including films, filmstrips, slides, audiotapes, and printed materials by category of science subject area.

EISENHOWER NATIONAL CLEARINGHOUSE FOR MATHEMATICS AND SCIENCE EDUCATION - 1994 GUIDEBOOK TO EXCELLENCE
Midwest Consortium for Mathematics and Science Educaion, 1900 Spring Rd, Suite 300, Oak Brook, IL 60521-1480 708-218-1268
A directory of federal resources for mathematics and science education improvement for the North Central Region. Copies are available while supplies last.

ENERGY EDUCATION RESOURCES - KINDERGARTEN THROUGH 12TH GRADE
National Energy Information Center, EI-231, Energy Information Administration, Room 1F-048, Forrestal Building, 1000 Independence Ave, SW, Washington, DC 20585 202-586-8800 1992
Ask for a copy of this 31 page booklet listing 86 different sources of educational materials from both public and private institutions and companies. Each source usually offers a catalog listing free materials.

ENVIRONMENTAL EDUCATION RESOURCE LIST
Contact Anne Marie Smith, Environmental Education Programs, Division of Educational Programs DEP/223, Argonne National Laboratory, 9700 S Cass Ave, Argonne, IL 60439-4845 708-252-7613
This extensive guidebook, available to teachers grades K-12+, contains over 200 references to environmental education resources in the Chicago area and nationwide.

EVERY TEACHER'S SCIENCE BOOKLIST
by the Museum of Science & Industry, compiled and edited by Bernice Richter and Pamela Nelson Scholastic Professional Books, Scholastic Inc., 2931 E McCarty St, Jefferson City, MO 65102 800-325-6149 1994 182 pages $ 18.95
This extensive bibliography of science literature for children lists and describes trade books by science topic. It also describes resource books for adults.

EXPLORING SCIENCE: A GUIDE TO CONTEMPORARY SCIENCE AND TECHNOLOGY MUSEUMS
Association of Science-Technology Centers, 1025 Vermont Ave, NW, Suite 500, Washington, DC 20005 202-783-7200 1980 72 pages In print
This is a complete, but older, directory to science and technology museums across the United States as well as affiliate international member museums.

ICE PICKS
by the Institute of Chemical Education, Department of Chemistry, University of Wisconsin-Madison, 1101 University Ave, Madison, WI 53706-1396 608-262-3033
This reference is a list of science activity books.

4 Science Fun in Chicagoland

IDEAAAS
edited by B. Walthall   The Learning Team, Suite 256, 10 Long Pond Rd, Armonk, NY 10504-0217   914-273-2226   1995   235 pages $ 24.95
This sourcebook for science, mathematics and technology education lists resources under major categories: Activity Guides, National Resources, State Resources, and Print and Non-print Resources.

INFORMATION SOURCES IN SCIENCE AND TECHNOLOGY - SECOND EDITION
by C. D. Hurt   Libraries Unlimited, Inc., P. O. Box 6633, Englewood, CO 80155-6633   800-237-6124   1994   350 pages $ 32.00   IBM Disks, $ 32.50
This bibliography lists reference and resource books and is in the Library Science Text Series. Resource books are listed under the following categories: History of Science, Multidisciplinary Sources, Astronomy, General Biology, Botany, Chemistry, Geosciences, Mathematics, Physics, Zoology, General Engineering, Civil Engineering, Energy & Environment, Mechanical & Electrical Engineering, Production Engineering, Transportation Engineering, and Biomedical Sciences.

NORTH CENTRAL REGIONAL EDUCATIONAL LABORATORY
1900 Spring Rd, Suite 300, Oak Brook, IL 60521-1480   800-356-2735
One of ten regional laboratories in the United States, NCREL disseminates information to schools and teachers about effective programs, develops educational products, holds conferences, and conducts research and evaluation. Ask for their catalog of publications available for all areas of education.

QUICK-SOURCE: THE DIRECTORY OF EDUCATIONAL TECHNOLOGY RESOURCES 1993-94 EDITION
AM Educational Publishing, Tallahassee, FL 32308   1993
This general resource computerized data base is also available in print.

RESOURCES FOR SCIENCE, MATHEMATICS, TECHNOLOGY AND EDUCATION
by the Teachers Academy for Mathematics and Science (TAMS), 10 W 35th St, Chicago, IL 60616   312-808-0100   8 pages
Names, addresses and telephone numbers are listed under two categories:
1) Science Centers, Museums, and Zoos   2) Institutions/Organizations Promoting Science, Mathematics and Technology Education.

RESOURCES FOR TEACHING GEOLOGY FROM THE ILLINOIS STATE GEOLOGICAL SURVEY
by the Illinois State Geological Survey, Natural Resources Building, 615 E Peabody Dr, Champaign, IL 61820   217-333-4747   1994   28 pages   Free
This catalog offers low-cost publications useful in teaching Illinois geology, geological maps, and information about Illinois geological science field trips. Also ask for brochure listing current Geological Science Field Trips in Illinois.

Chapter 01 - Books 5

Playing with these lighted optic fibers is a preschooler's way to learn about science wonders. Photo courtesy of Kohl Children's Museum, Wilmette, Illinois.

6 Science Fun in Chicagoland

RESOURCES FOR THE SCIENCE CLASSROOM
by Susan Dahl Fermi National Accelerator Laboratory, Leon M. Lederman Science Education Center, P. O. Box 500 MS 777, Batavia, IL 60510 708-840-8258 Fall, 1991.
A directory of over 150 sources for teachers of science. It was printed from a resource data base maintained at the Lederman Science Education Center. Reprinted every three years.

SCIENCE & TECHNOLOGY IN FACT AND FICTION: A GUIDE TO CHILDREN'S BOOKS
by Kennedy, Spangler and Vanderwerf R.R. Bowker, New York, 1990 319 pages
This guide to children's books is divided into categories of science and technology and subcategories of fiction and nonfiction. Each book listed is summarized and evaluated. This reference work includes indices author, title, subject and readability.

SCIENCE & TECHNOLOGY IN FACT AND FICTION: A GUIDE TO YOUNG ADULT BOOKS
by Kennedy, Spangler and Vanderwerf R.R. Bowker, New York, 1990 363 pages
This guide to young adult books is divided into categories of science and technology and subcategories of fiction and nonfiction. Each book listed is summarized and evaluated. This reference work includes indices by author, title, subject and readability.

SCIENCE 2001 TEXT SETS
Contact Dr. Diane Schiller, Loyola University Chicago, 6525 N Sheridan Rd, Chicago, IL 60626 312-508-8383 Age: Elementary School through High School
Teachers can borrow sets of science books grouped by subject, including inventors, insects, animals, space, weather, experimenting, water, earth science, states of matter, the egg, maple sugaring, human body, rainforests, science of music and spiders.

SCIENCE FOR CHILDREN: RESOURCES FOR TEACHERS
by the National Science Resources Center, Smithsonian Insitutution - National Academy of Sciences, National Academy Press, P. O. Box 285, Washington, DC 20055 800-624-6242 1988 177 pages $ 9.95
This extensive resource book lists and describes curriculum materials by science subject area, supplementary resources, and sources of information and assistance.

SCIENCE HELPER K-8 CD-ROM
The Learning Team, 10 Long Pond Rd, Armonk, NY 10504-0217 914-273-2226 800-793-TEAM $ 195.00
At the touch of a button, a teacher can have access to 919 lesson plans that include 2000 activities. Developed at the University of Florida under the direction of Dr. Mary Budd Rowe, this resource includes curriculum materials developed and tested over a 15-year period with millions of dollars in funding from the National Science Foundation.

SCIENCE SOURCES 1994
compiled by the Office of Communications, American Association for the Advancement of Science, 1333 H St N W, Washington, DC 20005 800-222-7809 1994 Published Annually 186 pages $ 15.00
This resource list includes the following categories: AAAS Sources, Colleges and Universities, Congressional Committees, Industrial Research, Federal Agencies and Laboratories, Hospitals and Medical Centers, International Science Sources, Museums of Science and Technology, Non-profit Research Institutions, Scientific Professional Societies, State Academies of Science as well as alphabetical and geographic indices.

A SCIENCE TEACHER'S DESK REFERENCE
prepared by Illinois State Board of Higher Education and Illinois State University and printed by the Illinois Science Teachers Association, University of Illinois, College of Education, 1310 S Sixth St, Champaign, IL 61820 217-244-0173 1993 24 pages Free
This booklet was authored by a group of ten Illinois science teachers to provide elementary (K-8) school teachers in Illinois with information and references useful for improving science instruction and assessment. Topics include Illinois Law, Assessment, the Four State Goals in Science, and Resources.

SCIENCE TRADEBOOKS, TEXTBOOKS AND TECHNOLOGY RESOURCES WITH ANNOTATIONS
Contact Fredric Tarnow, Science Coordinator, Lake County Educational Service Center, 19525 W Washington, Grayslake, IL 60030 708-223-3400 55 pages
This resource list includes over 700 Illinois Science Literacy Grant Publisher Gifts and Purchases. Listed alphabetically each resource is classified into subject area with a brief annotation.

Books on Activities and Methods of Teaching

These resources were selected from hundreds of similar publications available and are a list of resource books with exciting ideas for science learning activities.

175 AMAZING NATURE EXPERIMENTS
by Rosie Harlow and Gareth Morgan Distributed by the National Science Teachers Association 800-722-NSTA 1991 176 pages $ 12.00
This illustrated activity resource book covers How Things Grow, Minibeasts, Trees and Leaves, and The Seasons.

700 SCIENCE EXPERIMENTS FOR EVERYONE - REVISED AND ENLARGED
compiled by UNESCO Doubleday & Company, Inc., Garden City, New York 1956, 1962 250 pages
This classic reference lists activity experiments by categories of science subject area.

AMUSEMENT PARK PHYSICS: A TEACHER'S GUIDE
by Nathan A. Unterman J. Weston Walch, Publisher, 321 Valley St, P. O. Box 658, Portland, ME 04104-0658 800-341-6094 159 pages $ 19.95
This guide provides tutorials, practice problems, and lab exercises appropriate for studying the motion of amusement park rides.

ASTRONOMY FOR EVERY KID
by Janice VanCleave Distributed by the National Science Teachers Association 800-722-NSTA 1991 234 pages $ 10.95
This book contains 100 ideas, projects and activities for the classroom or science fair project.

CLASSROOM CREATURE CULTURE: ALGAE TO ANOLES - REVISED EDITION
by Carol Hampton, Carolyn Hampton, and David Kramer Published and distributed by the National Science Teachers Association 800-722-NSTA 1994 96 pages $ 12.95
Ideas for collecting, caring for, and investigating plants and simple animals in the science classroom grades K-9. This book is dedicated to understanding the respect and care living things need.

DEMONSTRATION EXPERIMENTS IN PHYSICS
edited by Richard Manliffe Sutton, Ph.D. prepared under the auspices of The American Association of Physics Teachers McGraw-Hill Book Company, Inc., New York 1938 545 pages Out of print and hard to find.
Perhaps the best compendium of classroom demonstrations for physics yet published.

DEMONSTRATION HANDBOOK FOR PHYSICS - SECOND EDITION
edited by Freier and Anderson American Association of Physics Teachers, One Physics Ellipse, College Park, MD 20740-3845 301-209-3300 320 pages $ 24.00
This handbook contains hundreds of apparatus demonstrations that require only low-cost, everyday materials.

EARLY CHILDHOOD AND SCIENCE
compiled by Margaret McIntyre Published and distributed by the National Science Teachers Association 800-722-NSTA 1984 136 pages $ 12.95
Skills of observation, identification and exploration are the focus of this book for the preschool child. It has application for the young gifted child as well as the slow learner.

EARTH SCIENCE FOR EVERY KID
by Janice VanCleave Distributed by the National Science Teachers Association 800-722-NSTA 1991 244 pages $ 10.95
This book contains 100 ideas, projects and activities for the classroom or science fair project.

THE EVERYDAY SCIENCE SOURCEBOOK
by Lawrence F. Lowery Distributed by the National Science Teachers Association 800-722-NSTA 1985 438 pages $ 21.00
This book is filled with activity ideas for teaching science in the elementary and middle school.

EXPERIMENTAL SCIENCE - VOLUMES I AND II
by George M. Hopkins Reprinted by Lindsay Publications, Inc., P. O. Box 538, Bradley, IL 60915-0538 815-935-5353 1902, 1987 Volume I 531 pages $ 19.95, Volume II 558 pages $ 19.95, Volumes I & II $ 34.95
These two books are packed with turn-of-the-century science activities on elementary and practical physics. Ask for 40 page catalog of Lindsay Publications. This publisher reprints many fine older books on science and technology.

10 Science Fun in Chicagoland

EXPERIMENTING WITH INVENTIONS
by Robert Gardner   Watts Publishing Co., New York   1990   128 pages $ 11.90
The process of inventing, the importance of the inventor's notebook, school invention clubs and fairs, and stories about inventors are topics in this book about inventing. This book is appropriate for adults, although it was written for junior high school students.

EXPERIMENTING WITH MODEL ROCKETS
Distributed by the National Science Teachers Association   800-722-NSTA   1989   86 pages $ 15.00
Teacher's guide. Explains what influences a model rocket's flight and how to measure the flight's height.

EXPLORATORIUM SCIENCE SNACKBOOK
Exploratorium To Go! Catalog, Exploratorium, 3601 Lyon St, San Francisco, CA 94123   800-359-9899   1991   450 pages $ 24.95   (800-722-NSTA $ 29.95)
This books shows how to build over 100 classroom interactive science exhibits like those found at the hands-on Exploratorium science center.

HELPING YOUR CHILD LEARN SCIENCE
U. S. Department of Education, Office of Educational Research & Improvement, Washington, DC 20208-5572 Order from: Superintendent of Documents, P. O. Box 371954, Pittsburgh, PA 15250-7954   202-783-3238   57 pages $ 3.25
This illustrated publication is full of activity ideas for the home and in the community. Science is described as observing, predicting and testing predictions. This book describes nine important concepts for science curriculum design recommended by the National Center for Improving Science Education.

INVENTING, INVENTIONS AND INVENTORS:
A TEACHING RESOURCE BOOK
by Jerry D. Flack   Teacher Ideas Press, A Division of Libraries Unlimited, Inc., P. O. Box 6633, Englewood, CO 80155-6633   800-237-6124   1989   148 pages $ 21.50
This book includes the process of inventing, methods of teaching, inventing competitions, and a resource bibliography.

INVITATIONS TO SCIENCE INQUIRY - SECOND EDITION
by Tik L. Leim   Science Inquiry Enterprises, 14358 Village View Lane, Chino Hills, CA 91709   909-590-4618
This popular book is filled with science teaching ideas.

Chapter 01 - Books 11

INVESTIGATING SCIENCE WITH DINOSAURS
by Craig A. Munsart Distributed by the National Science Teachers Association 800-722-NSTA 1993 250 pages $ 23.00
Using activities, worksheets and illustrations this book uses the popular interest in dinosaurs to introduce science. It contains a list of suggested resources.

KITCHEN SCIENCE
by Howard Hillman Exploratorium To Go! Catalog, Exploratorium, 3601 Lyon St, San Francisco, CA 94123 800-359-9899 $ 9.95
Learn about the science behind cooking and about the physics and chemistry in your kitchen. This fun book includes experiments and recipes.

MR. WIZARD'S SUPERMARKET SCIENCE
by Don Herbert, Television's Mr. Wizard Distributed by the National Science Teachers Association 800-722-NSTA 1980 96 pages $ 9.00
In this book you will learn that your local supermarket is a resource of everyday items that can be used in over 100 hands-on science experiments.

MULTICULTURALISM IN MATHEMATICS, SCIENCE AND TECHNOLOGY: READINGS AND ACTIVITIES
Distributed by the National Science Teachers Association 800-722-NSTA 1993 206 pages with wall poster $ 34.00
This book is filled with activities keyed to the activity's cultural origin. The enclosed wall poster shows a map of the world with pictures of famous scientists and inventors from all over the world.

PHYSICS BEGINS WITH AN M... MYSTERIES, MAGIC AND MYTH
by John W. Jewett, Jr. Longwood Professional Books, Allyn and Bacon, A Division of Simon & Schuster, Inc., Distributed by Paramount Publishing, P. O. Box 10695, Des Moines, IA 50336-0695 800-278-3525 1994 432 pages $ 37.95
This resource book describes physics activities that can be performed in the elementary through high school classroom using common, everyday apparatus.

PHYSICS EXPERIMENTS FOR CHILDREN
by Muriel & Mandell Dover Publications, 180 Varick St, New York, NY 1959 In print
One of a series of books with experiments and activities. Others include biology, chemistry, electricity, and astronomy.

PHYSICS OF SPORTS
edited by C. Frohlich American Association of Physics Teachers, One Physics Ellipse, College Park, MD 20740-3845 301-209-3300 124 pages $ 20.00
This collection of reprinted scientific journal articles presents the physics involved in today's baseball, basketball, bowling, golf, tennis, and track and field.

12 Science Fun in Chicagoland

PHYSICS OLYMPICS HANDBOOK
edited by Agrusco, Escobar and Moore American Association of Physics Teachers, One Physics Ellipse, College Park, MD 20740-3845 301-209-3300 26 pages $ 12.00
This handbook offers suggestions for organizing a physics olympics in your high school complete with rules, events, sample handouts, and schedules.

QUICK SCIENTIFIC TERMINOLOGY
by Kenneth Jon Rose John Wiley & Sons, Inc., New York 1988 267 pages
Learn scientific terminology with this self study guide by combining simple root words like megawatt, or mega + watt, means one million watts and creophagous, or creo + phagous, means flesh eating.

SCIENCE EXPERIENCES: COOPERATIVE LEARNING AND THE TEACHING OF SCIENCE
by Jack Hassard Iri Skylight, 200 E Wood St, Suite 274, Palatine, IL 60067 800-348-4474 $ 33.75 (800-722-NSTA $ 33.75)
This idea book for cooperative learning with science activities is available from two educational distributors, Iri Skylight and NSTA.

SCIENCE FAIRS AND PROJECTS - GRADES 7-12
Published and distributed by the National Science Teachers Association 800-722-NSTA 1988 72 pages $ 9.50
This book is for the teacher and explains all aspects of how to develop a successful science fair.

SCIENCE FAIRS AND PROJECTS - GRADES K-8
Published and distributed by the National Science Teachers Association 800-722-NSTA 1988 72 pages $ 9.50
This book is for the teacher and explains all aspects of how to develop a successful science fair.

SCIENCE IS... : A SOURCEBOOK OF FASCINATING FACTS, PROJECTS AND ACTIVITIES
by Susan V. Bosak Scholastic Canada Ltd., 123 Newkirk Road, Richmond Hill, Ontario, Canada L4C 3G5 416-883-5300 1991 515 pages $ 29.95 (1992 Edition 800-722-NSTA $ 29.95)
This extensive resource contains 489 pages of illustrated science activities listed by three categories: Quickies, Make Time, and One Leads to Another. It also has 18 pages listing over 200 other resource books.

SCIENCE ON A SHOESTRING - SECOND EDITION
by Herb Stongin Distributed by the National Science Teachers Association 800-722-NSTA 1991 208 pages $ 19.95
This book describes experiments with inexpensive materials. For grades K-8.

SCIENCE STARTERS
by Robert G. Hoehn   The Center for Applied Research in Education, a Divison of Simon & Schushter, Professional Publishing, West Nyack, New York 10995   1993
Over 1000 ready to use attention grabbers to make science fun, grades 6-12.

SCIENCEWORKS
by the Ontario Science Center   Distributed by the National Science Teachers Association   800-722-NSTA   1988   86 pages $ 9.95
Simple science experiments are presented in cartoon illustrations as fun mind-puzzling tricks. This book is a favorite of teachers and students.

STRING AND STICKY TAPE EXPERIMENTS
edited by Ronald Edge   American Association of Physics Teachers, One Physics Ellipse, College Park, MD 20740-3845   301-209-3300   448 pages $ 24.00
Describes simple experiments that may be constructed with materials readily available from a discount store.

TEACH THE MIND, TOUCH THE SPIRIT:
A GUIDE TO FOCUSED FIELD TRIPS
The Field Museum, Education Department, Roosevelt Road at Lake Shore Drive, Chicago, IL 60605-2497   312-922-2497, ext 351   80 pages $ 10.00
This book describes museums as educational opportunities, structuring your field trip, The Field Museum opportunities, and a reference bibliography.

TEACH YOUR CHILD SCIENCE:
MAKING SCIENCE FUN FOR THE BOTH OF YOU
by Michael Shermer   Contemporarty Books, Chicago   1989   149 pages $ 9.95
Part I, Getting Excited about Science, discusses science education. Part II, Doing Science, describes hands on activities and materials. Seven appendices include lists of resources.

TEN-MINUTE FIELD TRIPS
by Helen Ross Russell   Distributed by the National Science Teachers Association   800-722-NSTA   1991   176 pages $ 16.95
More than 200 short, close-to-home excursions in science for grades K-8 are described. Each excursion is categorized by science subject area and lists classroom activities with teacher preparation needs. Excursions are described for both rural and urban locations. Recommended for both teachers and parents.

14 Science Fun in Chicagoland

THOMAS EDISON BOOK OF EASY AND INCREDIBLE EXPERIMENTS
by James Cook and the Thomas Alva Edison Foundation Distributed by the National Science Teachers Association 800-722-NSTA 1988 136 pages $ 12.95
This book includes descriptions of the most popular experiments and projects sponsored by the Edison Foundation.

TOYS IN SPACE: EXPLORING SCIENCE WITH THE ASTRONAUTS
by Dr. Carolyn Sumners, Project Director for the Toys in Space Program, NASA TAB Books, Division of McGraw-Hill, Inc., Blue Ridge Summit, PA 17294-0840 800-822-8158 1994 141 pages Hard cover $ 17.95, paperback $ 10.95 (800-722-NSTA $ 10.95)
Many mechanical action toys were taken on a NASA shuttle mission to observe their motion in weightless space. These toys, how they move on Earth, and what happened to them in space are described.

Professional Books on Teaching Philosophy, Science History, and Science Reference

These books should be helpful references providing guidance, answers and encouragement to professional science teachers at all levels.

ASIMOV'S BIOGRAPHICAL ENCYCLOPEDIA OF SCIENCE AND TECHNOLOGY - SECOND EDITION
by Isaac Asimov Doubleday, New York, NY 1982 941 pages $ 29.95
This single volume chronologically describes the lives and achievements of over 1,000 great scientists from ancient times to the present.

ASIMOV'S CHRONOLOGY OF SCIENCE & DISCOVERY
by Isaac Asimov Harper & Row, Publishers, 10 E 53rd St, New York, NY 10022 1989 707 pages
The history of science from 4,000,000 B.C. to the present, listed by calendar year.

BIOLOGY TEACHER'S SURVIVAL GUIDE
by Michael F. Fleming Distributed by the National Science Teachers Association 800-722-NSTA 1993 304 pages $ 29.95
This book has listed tips, techniques and materials for success in the biology classroom.

CHICAGO PUBLIC SCHOOLS - CURRICULUM GUIDES
Contact Melanie Wojtulewicz, Department of Instructional Support, Chicago Public Schools, 1819 W Pershing Rd, Chicago, IL 60609   312-535-8850
The Department of Instructional Support has produced several curriculum guides including Science Grade 7, Science Grade 8, Model Science Supplement Grades 7-8 for magnet programs, Secondary School Life Science, Secondary School Physical Science, Secondary School Biology, Secondary School Chemistry, and Secondary School Physics.

CRC HANDBOOK OF CHEMISTRY AND PHYSICS
CRC Press, Inc., 2000 Corporate Blvd, NW, Boca Raton, FL 33431-9868   800-272-7737   Annually $ 99.50
This book often becomes the standard source for data and information about the physical properties of matter. It is available in most libraries.

DICTIONARY OF SCIENTIFIC BIOGRAPHY
edited by Charles C. Gillespie   Scribner, New York   1970-1980   16 volumes
This reference work can be found in major libraries and provides an excellect record of the history of science. It contains scholarly biographies of over 5,000 scientists from ancient to modern times.

EDTALK: WHAT WE KNOW ABOUT MATHEMATICS TEACHING AND LEARNING
North Central Regional Educational Laboratory (NCREL), 1900 Spring Rd, Suite 300, Oak Brook, IL 60521-1480   708-571-4700   800-356-2735   69 pages
A booklet for teachers and parents that answers some of the most frequently asked questions about the teaching and learning of mathematics.

EDTALK: WHAT WE KNOW ABOUT SCIENCE TEACHING AND LEARNING
North Central Regional Educational Laboratory (NCREL), 1900 Spring Rd, Suite 300, Oak Brook, IL 60521-1480   708-571-4700   800-356-2735   95 pages
A booklet for teachers and parents that answers some of the most frequently asked questions about the teaching and learning of science.

ELEMENTARY SCHOOL SCIENCE FOR THE 90'S
by Loucks-Horsley, Kapitan, Carlson, Kuerbis, Clark, Nelle, Sachse, and Walton   Association for Supervision and Curriculum Development, Alexandria, VA   166 pages $ 13.95
This work was supported by the U. S. Department of Education and the Office of Educational Research and Improvement. Some of the ideas presented in this book include Make Science a Basic, View Science Learning from a Constructivist Perspective, Assess What Is Valued, View Teacher Development as a Continuous Process, and Provide Teachers with Adequate Support to Implement Good Science Programs.

16 Science Fun in Chicagoland

MATHEMATICS, SCIENCE, AND TECHNOLOGY EDUCATION PROGRAMS THAT WORK
U. S. Department of Education, Office of Educational Research and Improvement Programs for the Improvement of Practice, National Diffusion Network, Washington, DC 20208-5645 (Illinois Statewide Facilitator 618-524-2664) 145 pages
A collection of 64 exemplary education programs and practices in the National Diffusion Network.

MC GRAW-HILL ENCYCLOPEDIA OF SCIENCE & TECHNOLOGY - 7TH EDITION
McGraw-Hill Inc., New York 1992 20 volumes
This excellent, international encyclopedia of science and technology covers topics alphabetically.

PROJECT 2061'S BENCHMARKS FOR SCIENCE LITERACY
Oxford University Press, 200 Madison Ave, New York, NY 10016 800-451-7556 1994 400 pages $ 35.00 includes computer disk with book.
A text of national standards of what all students should know and be able to do in natural and social sciences, mathematics, and technology by the end of grades 2, 5, 8, and 12. Project 2061, American Association for the Advancement of Science.

PROMISING PRACTICES IN MATHEMATICS & SCIENCE EDUCATION
North Central Regional Educational Laboratory (NCREL), 1900 Spring Rd, Suite 300, Oak Brook, IL 60521-1480 708-571-4700 800-356-2735 158 pages
This book describes a collection of promising educational programs and practices used in schools across the United States.

SCIENCE FARE
by Wendy Saul Harper & Row, Publishers, New York, 1986 Paperback, 295 pages
A classic illustrated guide and catalog of toys, books and activities about science for kids. The first nine chapters discuss science education for parents and teachers and the last eleven chapters describe specific resources by subject.

SCIENCE FOR ALL AMERICANS
by Rutherford Oxford University Press, 200 Madison Ave, New York, NY 10016 800-451-7556 1989 272 pages $ 11.95
Goals established in this work for national science education standards were used in preparation of *Project 2061's Benchmarks for Science Literacy*, a comprehensive list of what students should be able to do in science, mathematics, and technology by the AAAS.

These teachers are enjoying a summer workshop in the Golden Apple Science Program. Photo courtesy of the Golden Apple Foundation.

SCIENCE FOR THE ELEMENTARY SCHOOL - SEVENTH EDITION
by Victor and Kellough Macmillan Publishing Company, New York 1993
This college text is a compendium of ideas and knowledge about teaching elementary school science. See university bookstores in your area.

SCIENCE IN ELEMENTARY EDUCATION - FIFTH EDITION
by Peter C. Gega Macmillan Publishing Company, New York 1986
This text for college courses is a compendium of ideas and knowledge about teaching elementary school science. See university bookstores in your area.

SCIENCE THROUGH CHILDREN'S LITERATURE: AN INTEGRATED APPROACH
by Butzow & Butzow Teacher Ideas Press, A division of Libraries Unlimited, Inc., P. O. Box 6633, Englewood, CO 80155-6633 800-237-6124 1989 234 pages $ 24.50 (800-722-NSTA $ 24.50)
Unique primary school approach to science through children's stories.

TEACHING CHEMISTY TO STUDENTS WITH DISABILITIES - 3RD EDITION
American Chemical Society, Chicago Section, 7173 N Austin, Niles, IL 60714 708-647-8405 1993 46 pages Free
Ask for this excellent information and resource booklet.

TEACHING SCIENCE THROUGH DISCOVERY - SIXTH EDITION
by Carin and Sund Charles E. Merrill Publishing Company, Columbus, OH 1990
This college textbook for science teaching methods courses contains ideas and knowledge about teaching elementary school science. Contact university bookstores in your area.

THE GUIDE TO MATH & SCIENCE REFORM - AN INTERACTIVE RESOURCE FOR THE EDUCATION COMMUNITY
The Annenberg/CPB Math and Science Project, 901 E Street, NW, Washington, DC 20004 Publication office: Toby Levine Communications, Inc., 7910 Woodmont Ave, Suite 1304, Bethesda Maryland 20814 301-907-6510
Available in Macintosh and MS-DOS/Windows. This guide is an interactive computer data base of resources for the education community.

THE TIMETABLES OF SCIENCE
by Alexander Hellemans and Bryan Bunch A Touchstone Book, Published by Simon & Schuster Inc., Rockefeller Center, 1230 Avenue of the Americas, New York, NY 10020 1988 660 pages
This chronology of science lists science and its discoveries from 2,400,000 B.C. to present time by calendar year.

THE TRADITION OF SCIENCE
by Leonard C. Bruno Library of Congress, Washington, DC 1987 351 pages
The history of all fields of science are described with illustrations from early science books held in the collections of The Library of Congress.

Local Bookstores

These Chicagoland bookstores were selected for their exceptional offerings of science books.

Retail bookstores usually sell either new or used books. Bookstores selling new books provide a source for science in-print books currently available from their publisher. Bookstores that sell used books are fine sources for out-of-print books where one can find that unexpected treasure. If a certain title is desired, however, the buyer must hunt through a collection of used science books with little hope of finding it.

Reference book listing local bookstores:
THE BOOK LOVER'S GUIDE TO CHICAGOLAND
by Lane Phalen Brigadoon Bay Books, P. O. Box 957724, Hoffman Estates, IL 60195-7724 1992 256 pages $ 14.95
(Second edition available Spring, 1995)
This handy reference describes 400 bookstores in the Chicagoland area. Bookstores are categorized by geographic location and indexed by subject including science.

57TH STREET BOOKS
1301 E 57th St, Chicago, IL 60637 312-684-1300
New in-print books in all fields including science.

THE ADLER PLANETARIUM SHOP
Adler Planetarium, 1300 S Lake Shore Drive, Chicago, IL 60605 312-922-7827
This store within the Adler Planetarium carries a good selection of books on astronomy, astrophysics and space science.

B DALTON BOOKSELLER
175 W Jackson Blvd, Chicago 312-922-5219; 222 Merchandise Mart Plaza, Chicago 312-329-1881; 645 N Michigan Ave, Chicago 312-944-3702; 129 N Wabash, Chicago 312-236-7615. (See telephone yellow pages for store near you.)
Retail outlets for new in-print books.

20 Science Fun in Chicagoland

BARNES & NOBLE BOOKSTORES
1 S 550, Route 83, Oakbrook Terrace, IL 60181   708-571-0999; 659 W Diversey Ave, (just west of Clark St), Chicago, IL 60614   312-871-9004; 7 N Waukegan Rd (at Lakecook Rd), Deerfield, IL 60015   708-374-0320; 590 E Golf Rd, Schaumburg, IL 60173   708-310-0450; 351 Townsquare, Wheaton, IL 60187   708-653-2122; 1701 Sherman (at Church St), Evanston, IL 60201   708-328-0883; 5405 Touhy (Village Crossing Center), Skokie, IL 60077   708-329-8460.
Retail outlets for new in-print books.

BOOK STALL OF ROCKFORD
1032 Crosby St, Rockford, IL 61107 815-963-1671
General out-of-print books including technical books.

BOOKMAN'S CORNER
2959 N Clark St, Chicago, IL 60657 312-929-8298
Used out-of-print books in all fields including science.

BOOKSELLERS ROW
2445 N Lincoln Ave, Chicago, IL 60614 312-348-1170; 408 S Michigan Ave, Chicago, IL 60605 312-427-4242; and 1520 N Milwaukee Ave, Chicago, IL 60622 312-489-6200
General used and out-of-print books including a good selection of science books.

THE BOOKWORKS
3444 N Clark St, Chicago, IL 60657-1610 312-871-5318
General used and out-of-print books in all subject areas including science.

BORDERS BOOKS AND MUSIC
830 N Michigan Ave, Chicago, IL 60611   312-573-0564; 49 S Waukegan Ave, Deerfield, IL 60015   708-559-1999; 336 S Route 59, Naperville, IL 60540   708-637-9700; 1500 16th St, Oak Brook, IL 60521 708-574-0800; 1540 Golf Rd, Schaumburg, IL 60173 708-330-0031; 3232 Lake Ave, Wilmette, IL 60091 708-256-3220
New in-print books in all subject areas including science.

BROOKFIELD ZOO - BOOKSTORE
3300 S Golf Rd, Brookfield, IL 60513 708-485-0263
The Brookfield Zoo Bookstore and Shop offers more than 5,000 titles on zoology and natural history topics.

THE CHILDREN'S BOOKSTORE
2465 N Lincoln Ave, Chicago, IL 60614 312-248-2665
This bookstore carries new books for children including a good science selection. Inquire about scheduling a field trip, storyhour, book fair, or visit from the bookmobile. Ask for a school fundraising catalog written by children and parents.

DAN BEHNKE, BOOKSELLER
2463 N Lincoln, Chicago, IL 60614 312-404-0403
General used and out-of-print books including a modest science selection.

EVANSTON ECOLOGY CENTER - BOOKSTORE
Evanston Environmental Association, Evanston Ecology Center, 2024 McCormick Blvd, Evanston, IL 60201 708-864-5181
At this bookstore in the Ecology Center you will find birdfeeding supplies, gardening books, puppets, children's books.

FIELD MUSEUM OF NATURAL HISTORY - BOOKSHOP
The Field Museum, Roosevelt Road at Lake Shore Drive, Chicago, IL 60605-2497 312-922-9410, ext 693
This bookshop and gift shop contains a large selection of quality books about natural science and culture.

GREAT EXPECTATIONS
911 Foster St, Evanston, IL 60201 708-864-3881
New in-print books on all subject areas including science.

HAMILL & BARKER
1719 Howard St, Evanston, IL 60202 708-475-1724
Bookshop open by appointment. Rare books in all fields including science and technology.

HELENA SZEPE, BOOKS
1525 E 53rd St, Hyde Park Bank Bldg, Suite 825, Chicago, IL 60615 312-493-4470
By appointment only. Rare books in all fields including science.

ILLINOIS INSTITUTE OF TECHNOLOGY BOOKSHOP
3200 S Wabash Ave, Chicago, IL 60616 312-567-3120
This university bookstore has new books in-print on science and technology.

22 Science Fun in Chicagoland

KROCH'S & BRENTANO'S INC.
29 S Wabash Ave, Chicago, IL 60603 312-332-7500; 1530 E 53rd, Chicago, IL 312-288-0145; 30 N LaSalle, Chicago, IL 312-704-0287; 516 N Michigan, Chicago, IL 312-321-0989; 230 S Clark St, Chicago, IL 312-553-0171; 2070 N Clyborn Ave, Chicago, IL 312-525-2800; 835 N Michigan, Chicago, IL 312-943-2452. (For suburban locations see telephone yellow pages.) Technical Book Department 312-332-7500.
New in-print books in all fields including science.

MUSEUM OF SCIENCE & INDUSTRY BOOKSTORE & SHOP
Museum of Science & Industry, 57th St and Lake Shore Drive, Chicago, IL 60637 312-684-1414, ext 2780
This retail store has its largest display at the main entrance within the Museum of Science & Industry. This store exclusively sells science books, toys and novelties.

N. FAGIN BOOKS
1039 W Grand Ave, Chicago, IL 60622 312-829-5252
New and used out-of-print books on anthropology, archeology, botany and zoology.

O'GARA & WILSON BOOKSELLERS, LTD.
1311 E 57th St, Chicago, IL 312-363-0993 (Second location 101 Broadway, Chesterton, IN 216-926-2066)
Established in 1882, this is Chicago's oldest bookstore. General used and out-of-print books including science.

POWELL'S BOOKSTORE
1501 E 57th St, Chicago, IL 60637 312-955-7780; 828 S Wabash, Chicago, IL 60605 312-341-0748; and 2850 N Lincoln, Chicago, IL 60657 312-248-1444
General used out-of-print books including scholarly books in all fields with a very good science selection.

RAIN DOG BOOKS
827 Foster, Evanston, IL 60201 708-869-2665
Collectable books in all fields including a modest science collection.

RICHARD ADAMIAK
1700 E 56th St, Chicago, IL 60637-1970 312-955-4571
By appointment only. General used and rare books including science.

SEMINARY CO-OP BOOKSTORE
5757 S University Ave, Chicago, IL 60637 312-752-4381
New in-print books on all fields including science.

THE STARS OUR DESTINATION
1021 W Belmont, Chicago, IL 60657 312-871-2722
New and used books on science fiction. The largest used science fiction selection in Chicago.

THE TIME MUSEUM STORE
Clock Tower Inn, 7801 E State St, (Interstate 90 and business highway 20), Rockford, IL 61125 815-398-6000
The Time Museum is an extraordinary museum of timekeeping, from Stonehenge to the Atomic clock. The store within the museum offers a fine selection of books on horology, astronomy, physics and mathematics.

U. S. GOVERNMENT BOOKSTORE
401 S State, Chicago, IL 60605 312-353-5133
This bookstore sells books printed by agencies of the U. S. Government.

UNIVERSITY OF CHICAGO BOOKSTORE
970 E 58th St, Chicago, IL 60637 312-702-8729
This university bookstore sells new in-print books on all fields including science.

UNIVERSITY OF ILLINOIS AT CHICAGO BOOKSTORE
750 S Halsted, Chicago, IL 60607 312-413-5500
This university bookstore sells new in-print books on all fields including science.

Mail Order Books For Sale

Where indicated some of these mail order sources sell used and out-of-print science books. Each resource provides a catalog of books for sale.

AAAS PRESS
American Association for the Advancement of Science, 1333 H St, NW, Washington, DC 20005 800-222-7809
Ask for catalog of publications. Ordering address: AAAS Distribution Center, P. O. Box 521, Annapolis Junction, MD 20701

AMERICAN ASSOCIATION OF PHYSICS TEACHERS - AAPT PRODUCTS CATALOG
American Association of Physics Teachers, One Physics Ellipse, College Park, MD 20740-3845 301-209-3300
Ask for this 20-page catalog. The 1994-95 catalog lists 65 books on physics and physics education. It also offers Physics of Technology Modules, computer software, videodiscs, videotapes, slide sets, materials for teachers, T-Shirts, workshop materials, and posters.

24 Science Fun in Chicagoland

AMERICAN GEOLOGICAL INSTITUTE
4220 King St, Alexandria, VA 22302 703-379-2480 AGI Publications Center, P. O. Box 205, Annapolis Junction, MD 20701 301-953-1744
Ask for 6-page brochure listing publications available, including Earth Science Guidelines Grades K-12, Earth-Science Education Resource Directory, and Careers in the Geosciences.

CRITICAL THINKING PRESS & SOFTWARE
P. O. Box 448, Pacific Grove, CA 93950 800-458-4849
Ask for 50-page catalog of books on thinking and teaching skills, including two pages specifically on science.

DALE SEYMOUR PUBLICATIONS
P. O. Box 10888, Palo Alto, CA 94303-0879 800-872-1100
Ask for 160-page catalog offering K-8 educational materials in mathematics and science.

EUREKA!
Lawrence Hall of Science, University of California, Berkeley, CA 94720 510-642-1016
Ask for 30-page Lawrence Hall of Science publications catalog, *Eureka!*, filled with books, curriculum materials, science kits and videos.

EXPLORATORIUM STORE
3601 Lyon St, San Francisco, CA 94123 415-561-0393 800-359-9899
Ask for the 30-page *Exploratorium To Go! Catalog* that is filled with quality science toys and books.

IDEA FACTORY, INC.
10710 Dixon Dr, Riverview, FL 33569 800-331-6204 Age: Grades K-8
Elementary and middle school science and math resources. Ask for 16-page catalog filled with books, science teaching ideas, and materials.

INTERNATIONAL TECHNOLOGY EDUCATION ASSOCIATION
1914 Association Dr, Reston, VA 22091-1502 703-860-2100
Ask for 20-page catalog of publications and classroom materials.

J. WESTON WALCH, PUBLISHER - SCIENCE CATALOG
J. Weston Walch, Publisher, 321 Valley St, P. O. Box 658, Portland, ME 04104-0658 800-341-6094
Ask for this 30-page catalog offfering books for science teachers. Many of these books were written by teachers for teachers.

KEY CURRICULUM PRESS - SECONDARY MATH CATALOG
Key Curriculum Press, 2512 Martin Luther King Jr. Way, P. O. Box 2304, Berkeley, CA 94702 800-338-7638
Ask for 20-page catalog of materials "advocating change though class action."

KNOLLWOOD BOOKS
P. O. Box 197, Oregon, WI 53575-0197 608-835-8861
Knollwood Books sells new, used and rare books on science, astronomy, meteorology, space and rocketry by mail order and at book fairs. Ask for a catalog.

LIBRARIES UNLIMITED/ TEACHER IDEAS PRESS
P. O. Box 6633, Englewood, CO 80155-6633 800-237-6124
Ask for the 62-page catalog listing reference books and teaching ideas books. Many books by this publisher are about science.

LITTLE MINDS: BOOKS FOR KIDS
P. O. Box 1038, Highland Park, IL 60035 800-574-3388
Ask for catalog of books offered, including science books.

NATIONAL ACADEMY PRESS
2101 Constitution Ave, NW, P. O. Box 285, Washington, DC 20055 800-624-6242
Ask for this publisher's catalog. The National Academy Press was created by the National Academy of Sciences. Its 39-page catalog lists publications about science and technology under twelve different subject categories.

NATIONAL ASSOCIATION OF BIOLOGY TEACHERS
11250 Roger Bacon Dr, #19, Reston, VA 22090-5202 703-471-1134
Ask for publications and membership brochure that includes *Favorite Labs from Outstanding Teachers*, *Biology Labs that Work: The Best of How-To-Do-Its*, and *Careers in Biology: An Introduction*.

NATIONAL COUNCIL OF TEACHERS IN MATHEMATICS (NCTM) - EDUCATIONAL MATERIALS CATALOG
National Council of Teachers in Mathematics, 1906 Association Dr, Reston, VA 22091-1593 800-235-7566
Ask for this 32-page catalog filled with quality books on the teaching of mathematics.

NATIONAL ENERGY FOUNDATION
5160 Wiley Post Way, Suite 200, Salt Lake City, UT 84116 801-539-1406
Ask for 15-page catalog of publications and science kits. Materials include *Out of the Rock*, a mineral resource and mining education program for K-8 produced in conjunction with the U. S. Bureau of Mines.

NSTA MEMBERSHIP & PUBLICATIONS SUPPLEMENT
to Science & Children and The Science Teacher
National Science Teachers Association, 1840 Wilson Blvd, Arlington, VA 22201-3000 800-722-NSTA Published annually Over 50 pages Free
Ask for this catalog. In 1995 the NSTA publications catalog offered over 180 quality books for sale for teachers as well as posters and NSTA membership information. Both parents and teachers will find this the single best source for science activity books.

ORYX
4041 N Central Ave, Suite 700, Phoenix, AZ 85012-3397 800-279-6799
Ask for this publisher's 62-page catalog of information books and guide books including 23 about science and technology.

PUBLICATIONS FOR PARENTS FROM THE U.S. DEPARTMENT OF EDUCATION
U.S. Department of Education, Education Information Branch, 555 New Jersey Ave, NW, Washington, DC 20208-5641
Ask for this brochure of publications, including *Helping Your Child Learn Math* and *Helping Your Child Learn Science.*

THINKING WORKS
P. O. Box 468, St. Augustine, FL 32085-0468 800-633-3742
Ask for 48-page catalog of books on teaching and thinking, including 4 pages on science.

UNITED STATES GOVERNMENT PRINTING OFFICE
Public Documents Distribution Center, Pueblo, CO 81009
Request a catalog of current publications or visit the U. S. Government Bookstore at 401 S State, Chicago, IL 60605, 312-353-5133.

Chapter 02
Computers

Computer Information Sources

If you are interested in learning about computers, the following list should be very helpful.

BOARDWATCH MAGAZINE:
GUIDE TO ELECTRONIC BULLETIN BOARDS AND THE INTERNET
8500 W Bowles Ave, Suite 210, Littleton, CO 80123   800-933-6038   Monthly
$ 36.00 per year. $ 4.95 per issue.
This 150-page magazine covers current news about linking to electronic bulletin boards and the Internet. Available at retail stores selling magazines.

28 Science Fun in Chicagoland

CD-ROM WORLD
Meckler Corporation, 11 Ferry Lane West, Westport, CT 06880-5880 203-226-6967 Monthly $ 29.00 per year.
Information on CD-ROM products for home and professional use.

CITY COLLEGES OF CHICAGO
Daley College 312-838-7500, Dawson Technical Institute 312-451-2000, Kennedy-King College 312-602-5000, Lakeview Learning Center 312-907-4400, Malcom X College 312-850-7000, Olive-Harvey College 312-568-3700, South Chicago Learning Center 312-291-6770, Truman College 312-878-1700, Washington College 312-553-5600, West Side Learning Center 312-850-7420, Wright College-North 312-777-7900, Wright College-South 312-481-8000.
Courses on computer training are available at most sites of the City Colleges of Chicago and at many community colleges in suburban areas. Inquire about current courses being offered usually at very reasonable fees.

COMPUTE
Compute Publications, International Ltd., 1965 Broadway, New York, NY 10023-5965 800-727-6937 Monthly $ 19.95 per year.
Each issue contains feature articles, columns, multimedia articles, reviews, and an entertainment section.

COMPUTER CHRONICLES
800-799-4949
This program is broadcast on WYCC/Chicago Channel 20. It summarizes current advances in personal computers and software. Videocassettes of individual programs and a newsletter subscription are offered at $ 32.50 each.

THE COMPUTING TEACHER
International Society for Technology Education, 1787 Agate St, Eugene, OR 97403-1923 503-346-4414
This professional organization is dedicated to the improvement of all levels of education through the use of computer-based technology.

EDUCATORS GUIDE TO FREE COMPUTER MATERIALS
Educators Progress Service, Inc., 214 Center St, Randolph, WI 53956-1497 414-326-3126 1994 $ 37.95
This book lists and describes free computer materials available to educators.

I.C.E. CUBE
Computer Update Bulletin for Educators (CUBE), Illinois Cumputing Educators (I.C.E.), 8548 145th St, Orland Park, IL 60462-2839 Contact Vicki Logan at 708-894-8574 Bimonthly $ 25.00 per year, includes membership.
This organization focuses on utilizing computer technology in the classroom. This newletter includes information about computer bulletin boards, information on grants, reviews of software, announcements about meeting where public domain software is traded.

MAC/CHICAGO
Peregrine Marketing Associates, Inc., 515 E Golf Rd, Suite 201, Arlington Heights, IL 60005 708-439-6575 Bimonthly $ 2.50 per issue.
Filled with resources, it is "The Resource for Chicagoland Macintosh Users."

MAC-USER
MacUser, 950 Tower Lane, 18th Floor, Foster City, CA 94404 415-378-5600 800-627-2247 Monthly Single copy $ 2.95. $ 27.00 per year.
This 200-page magazine is dedicated to the Apple computer and includes news, reviews, and advertisements. Available where magazines are sold.

MACWORLD
Macworld Communications, 501 Second St, San Francisco, CA 94107 Editorial, 415-243-0505 Subscriptions, 800-288-6848 Monthly Single issue $ 3.95. $ 30.00 per year.
This 250-page magazine is dedicated to Apple computers. Macworld is a publication of Macworld Communications and is an independent journal not affiliated with Apple Computer, Inc.

MEDIA AND METHODS MAGAZINE
1429 Walnut St, Philadelphia, PA 19102 215-563-6005 5 issues per year. $ 29.00 per year.
Magazine about technology in the classroom.

NSTA SCIENCE EDUCATION SUPPLIERS
A Supplement to: Science & Children, Science Scope, and The Science Teacher, National Science Teachers Association, 1840 Wilson Blvd, Arlington, VA 22201-3000 800-722-NSTA Published annually. 122 pages $ 5.00 per copy.
List of science educational software developers and distributors. The most current and comprehensive list of manufacturers, publishers and distributors of science education materials. See Computer Software.

30 Science Fun in Chicagoland

PC MAGAZINE:
THE INDEPENDENT GUIDE TO PERSONAL COMPUTING
Ziff-Davis Publishing Company, L.P., One Park Ave, New York, NY 10016-5802 Bimonthly Single copy $ 3.95. $ 49.97 per year.
This 400 page magazine is dedicated to all aspects of personal computing. Available where magazines are sold.

PC WORLD
PC World Communications, Inc., 501 Second St, # 600, San Francisco, CA 94107 415-243-0500 Bimonthly Single issue $ 3.95. $ 29.90 per year.
This 350 page magazine is dedicated to news about new products for home computer needs and includes numerous reviews and resources.

SCIENCE & TECHNOLOGY INFORMATION SYSTEM (STIS)
National Science Foundation, 4201 Wilson Blvd, Washington, DC 22230 Voice 703-306-0214, Modem 7E1F 703-306-0212 or 703-306-0213
STIS is an electronic dissemination system that provides fast, easy access to NSF publications. There is no cost to you except for possible long-distance phone charges. The service is available 24 hours a day. Science education programs receive funding through the Division of Undergraduate Education, Directorate for Education and Human Resources, National Science Foundation.

TECHNOLOGY AND LEARNING MAGAZINE
Available from the International Society for Technology Education, 1787 Agate St, Eugene, OR 97403-1923 503-346-4414 Published by Peter Li Education Group, Dayton, OH 45439-1597 513-847-5900 8 issues per year. $ 19.95 per year.
The professional organization, ISTE, is dedicated to the improvement of all levels of education through the use of computer-based technology. This magazine is about technology in the classroom.

T.H.E. JOURNAL (TECHNOLOGY IN EDUCATION JOURNAL)
Circulation Department, T.H.E. Journal, 150 El Camino Real, Suite 112, Tusin, CA 92680-3670
Request application for free subscription.

THE TECHNOLOGY TEACHER
International Technology Education Association, 1914 Association Dr, Reston, VA 22091 703-860-2100 Monthly $ 55 per year.
Each issue provides ideas for the classroom, project activities, resources in technology, and current trends in technology education.

The Internet

COMPUTING ENGINEERS INC.
P. O. Box 285, Vernon Hills, IL 60061-0285 Voice 708-367-1870, Modem 8N1 312-282-8605, Modem 8N1F 708-367-1871, telnet gagme.wwa.com Distributor of World-Wide Access. Telnet account $ 15.00 per month, $ 35.00 per quarter. Full Internet account $ 19.50 per month, $ 115.00 per half year, $ 220.00 per year.
WorldWide Access can be your personal link to the information superhighway known as the Internet. The Internet includes:

Electronic Mail (E-mail), a popular way to communicate on the Internet;

Usenet (or Netnews), a collection of 6,000 discussion groups where discussions take place about every imaginable subject;

Telnet, a command that allows you to log in to a computer system on the Internet and be able to use it directly;

FTP is a protocol and a program that allows you to transfer files to and from other computers on the Internet;

IRC (Internet Relay Chat) provides a system where many people may participate in a live computer-generated conversation;

WWW (World Wide Web) is a hypermedia information retrieval method to provide simple access to a huge body of information;

SLIP/PPP are two popular ways to send and receive mail directly from the Internet;

Commerical World-Wide Web/Gopher Services uses World-Wide Access or Gopher server space to provide organizations large and small with Internet access.

Computing Engineers Inc. is the parent company of World-Wide Access and is staffed by a team with many years of programming and networking experience. Contact Computing Engineers Inc. for current and further pricing information.

ILLINOIS STATE BOARD OF EDUCATION - INTERNET ACCESS
Contact Richard Dehart or Penny Kelly, Illinois State Board of Education, S-395 100 N First St, Springfield, IL 62777 217-782-2826
Ask for letter of request to obtain access to the Internet via the ISBE host site.

32 Science Fun in Chicagoland

AIR POLUTION BBS
telnet ttnbbs.rtpnc.epa.gov

CENTER FOR ADVANCED SPACE STUDIES
telnet cass.jse.nasa.gov   Login: cass   PW: online

FEDERAL GOVERNMENT DATABASES
telnet fedworld.doc.gov

GLOBAL LAND INFORMATION SYSTEMS
telnet glis.cr.usgs.gov   Login: guest

NEWTON (ARGONNE NATIONAL LABORATORY DIVISION OF EDUCATIONAL PROGRAMS BBS)
telnet newton.dep.anl.gov   Login: coconet (free access software required) or cocotext

NOAA (NATIONAL OCEANOGRAPHIC & ATMOSPHERIC ADMINISTRATION)
telnet esdim1.nodc.noaa.gov   Login: noaadir

SCIENCE AND TECHNOLOGY INFORMATION SYSTEMS
telnet stis.nsf.gov   Login: public

SPACEMET (NASA SCIENCE BBS)
telnet spacemet.phast.umass.edu

STINFO (HUBBLE TELESCOPE NEWS)
telnet stinfo.hq.eso.org   Login: stinfo

Computers and Education

Several interesting science education computer applications, groups, and opportunities are listed below:

ADAM SOFTWARE, INC.
1600 Riveredge Pkwy, Suite 800, Atlanta, GA 30328   800-755-ADAM
Ask for information about the *Animated Dissection of Anatomy for Medicine* scholar series. This multimedia application helps classrooms discover anatomy with dissections via computer.

AURBACH & ASSOCIATES, INC.
8233 Tulane Ave, St. Louis, MO 63132-5019 800-77G-RADY
Offers the Grady student portfolio assessment program for the Macintosh. Ask for a free demonstration disk.

THE GUIDE TO MATH & SCIENCE REFORM - AN INTERACTIVE RESOURCE FOR THE EDUCATION COMMUNITY
The Annenberg/CPB Math and Science Project, 901 E Street, NW, Washington, DC 20004 Publication office: Toby Levine Communications, Inc., 7910 Woodmont Ave, Suite 1304, Bethesda Maryland 20814 301-907-6510
Available in Macintosh and MS-DOS/Windows, this guide is an interactive computer data base of resources for the education community.

I.C.E. CUBE
Computer Update Bulletin for Educators (CUBE), Illinois Computing Educators (I.C.E.), 8548 145th St, Orland Park, IL 60462-2839 Contact Vicki Logan at 708-894-8574 Bimonthly $ 25.00 per year, includes membership.
This organization focuses on utilizing computer technology in the classroom. This newletter includes information about computer bulletin boards, information on grants, reviews of software, and announcements about meetings where public domain software is traded.

ILLINOIS COMPUTING EDUCATORS
Affiliate of International Society for Technology in Education, c/o Vicki Logan, 137 Foster, Roselle, IL 60172 708-960-1137 General membership $ 25.00 per year.
Membership includes programs, meetings, newsletter, network for resources, and Public Domain software. Various chapters of this organization meet in the Chicagoland area. Request a membership application.

INTERACTIVE PHYSICS
Knowledge Revolution, Prentice-Hall, Inc., Englewood Cliffs, NJ 07632 1995 269 pages Software disks included.
This interactive physics software with manual for students creates visual motion images and graphical analysis as the student selects the physical setting and parameters. This computer software has become the standard for modeling the physical world on the computer.

LAP TOP COMPUTER EDUCATION
Contact Bryan Wunar, Loyola University Chicago, 6525 N Sheridan Rd, Chicago, IL 60626 312-508-8383
Edgewater, Uptown, and Rogers Park schools without computers can have lap top computers brought to their classroom for a fun hands-on introduction to computers.

34 Science Fun in Chicagoland

LEON M. LEDERMAN SCIENCE EDUCATION CENTER - MATH SCIENCE CONSORTIA DEMONSTRATION SITE
Fermi National Accelerator Laboratory, P. O. Box 500 MS 777, Batavia, IL 60510   708-840-8258
At this science education center the visitor may preview the Internet, science education computer bulletin boards, and see multimedia materials. The Teacher Resource Center is a demonstration site for mathematics and science materials in the Math Science Consortia, North Central Regional Educational Laboratory for the Eisenhower National Clearinghouse. Phone for an appointment.

NEWTON
Contact Lou Harnisch, Educational Programs Coordinator   Division of Educational Programs DEP/223, Argonne National Laboratory, 9700 S Cass Ave, Argonne, IL 60439-4845   Voice 708-252-6925, telnet newton.dep.anl.gov, Modem 8N1F 708-252-8241
NEWTON is a free nationwide electronic bulletin board for science, math, computer science and technology teachers and students at any level. Features teacher idea exchange and ask a scientist. To access NEWTON with your computer and modem set your communications software to 8 bits, no parity, 1 stop bit, 2400-14,400 baud, full duplex and dial 708-252-8241. Type in "bbs" at the login prompt, follow the return prompts, select #1 on the menu and type "new", enter VT100, and proceed to register. Complete all information in lower case. NEWTON supports 32 simultaneous users, allowing Group Chat, and has registered over 9000 users. Teachers can use the Internet E-Mail and other features of the Internet (To apply see the Group Discussion called Sysop-System-ExtendFeat). Ask-A-Scientist and Discussion Topics are very popular NEWTON features.

PRODIGY
Prodigy Services Company, 445 Hamilton Ave, H8B, White Plains, NY 10601   800-PRODIGY   800-776-3449, ext 176
Ask for information brochure about the *Classroom Prodigy Offer* offering teleconnecting a classroom. Includes one flat annual fee, the Internet E-mail, Internet Digest, bulletin boards, teacher's manual with lesson plans, projects that link up with classrooms all over the country.

PROJECT INFORM
Contact Sarai Lastra, Chicago Public Schools Voice 312-535-4981, Modem 8E1F 312-535-8513, telnet cpsnet2.cps.edu
This electronic bulletin board for teachers and students is called CPSnet. Resources include ERIC, the largest educational database, The Chicago Tribune, Financial Aid Library, and college and school financial facts. Ask for a copy of the *Project Inform Reference Manual*.

SCIENCE HELPER K-8 CD-ROM
The Learning Team, 10 Long Pond Rd, Armonk, NY 10504-0217 914-273-2226 800-793-TEAM $ 195.00
At the touch of a button, a teacher can have access to 919 lesson plans that includes 2000 activities. Developed at the University of Florida under the direction of Dr. Mary Budd Rowe, this resource lists curriculum materials developed and tested over a 15-year period with millions of dollars in funding from the National Science Foundation.

SYNAPSYS SOFTWARE
8400 E Prentice Ave, Penthouse, Englewood, CO 80111   303-222-9022
Offers a windows based student assessment software package. Ask for brochure about application.

Local Computer and Software Suppliers

These retail stores compiled from computer educators' recommendations will help you when shopping for your computer.

BEST BUY
Chicago (North Ave & Sheffield) 312-988-4067, Betford Park 708-563-9080 (See telephone yellow pages for a store near you.)
Retail source for both hardware and software. Inquire about software training classes.

CHICAGO COMPUTER EXCHANGE
5225 S Harper Ave (Lower Level of Harper Court Shopping Center), Chicago, IL 60615   312-667-5221
For hardware needs this store buys and sells new or used computers, parts, and peripherals.

CIRCUIT CITY
2500 N Elston Ave, Chicago 312-772-0037   (See telephone yellow pages for a store near you.)
Retail source for both hardware and software.

COMPUSA
Skokie 708-933-4700, Schaumburg 708-619-1221, Downers Grove 708-241-1144
Retail source for both hardware and software. Ask to be placed on mailing list.

36 Science Fun in Chicagoland

COMPUTER DISCOUNT WAREHOUSE
Chicago 312-527-2700, Northbrook 708-564-4900 (See your telephone yellow pages under Computers-Dealers)
Retail source for both hardware and software. Ask to be placed on mailing list.

COMPUTERLAND
(See your telephone yellow pages under Computers)
Retail source for both hardware and software.

Multimedia interactive computer workstations at the Motorola Museum of Electronics encourage visitors to explore topics ranging from electricity to the binary system. Photo courtesy of the Motorola Museum of Electronics.

EGG HEAD SOFTWARE
Five Chicagoland locations.
(See your telephone directory under Egg Head Software)
Retail source for computer software.

Chapter 02 - Computers 37

ELEKTEK
6557 N Lincoln Ave, Lincolnwood, IL 60645   708-677-7660, ext 1100   (See your telephone yellow pages under Computers)
Retail source for both hardware and software. Ask to be placed on mailing list.

MONTGOMERY WARD ELECTRIC AVENUE
Chicago 4620 S Damen 312-650-9011, 2939 W Addison 312-509-1211, 7601 S Cicero 312-284-6414, 14736 S Campbell 708-371-9588   (See your telephone yellow pages under Appliances)
Retail source for both hardware and software.

OFFICE DEPOT
Lincoln Park 312-587-0863, Lincolnwood 312-583-5301 (See your telephone directory under Office Depot Inc.)
Retail source for both hardware and software.

OFFICEMAX
Nineteen Chicagoland locations. (See your telephone directory under OFFICEMAX)
Retail source for both hardware and software.

SEARS ROEBUCK AND CO.
(See your telephone directory under Sears)
Retail source for both hardware and software.

Mail Order Suppliers

Computer educators recommended these special mail order opportunities:

AMERICAN ASSOCIATION OF PHYSICS TEACHERS - AAPT PRODUCTS CATALOG
American Association of Physics Teachers, One Physics Ellipse, College Park, MD 20740-3845   301-209-3300
Ask for this 20-page catalog. The 1994-95 catalog lists 65 books on physics and physics education. It also offers *Physics of Technology Modules*, computer software, videodiscs, videotapes, slide sets, materials for teachers, T-Shirts, workshop materials, and posters.

APPLE DIRECT - EDUCATOR ADVANTAGE PROGRAM
P. O. Box 898, Lakewood, NJ 087019930   800-959-APPL
Ask for price lists for software and computers at a significant educator discount.

38 Science Fun in Chicagoland

BRODERBUND SOFTWARE
P. O. Box 6125, Novato, CA 94948-6125   800-521-6263   415-382-4400
Ask for 20-page catalog of this software distributor. Source of *Where in the World is Carmen Sandiego?*, and interactive, antimated stories for children.

DAVIDSON & ASSOCIATES
P. O. Box 2961, Torrance, CA 90509 800-545-7677
Ask for 50-page *School Catalog* listing a variety of educational software, CD-ROM's, and teacher support materials.

DEMCO
P. O. Box 7488, Madison, WI 53791-9955   800-356-1200
Ask for 800-page school supply catalog that contains 32 pages of computer supplies for education.

EDUCATIONAL ACTIVITIES, INC.
P. O. Box 392, Freeport, NY 11520 800-645-3739
See 70-page catalog of software, CD-ROM, videos, and multimedia for all ages.

EDUCATIONAL RESOURCES
1550 Executive Dr, Elgin, IL 60123 800-624-2926
Ask for catalog from this software distributor.

EDUCORP
7434 Trade St, San Diego, CA 92121   800-843-9497
Ask for catalog from this software distributor.

THE EDUTAINMENT CATALOG
932 Walnut St, Louisville, CO 80027 800-338-3844
Ask for 48-page catalog of educational software for children K-12, including DOS, Windows, Apple, and Mac formats listed for disc and CD-ROM.

INTERNATIONAL SOCIETY FOR TECHNOLOGY IN EDUCATION
1787 Agate St, Eugene, OR 97403-1923 800-336-5191   503-346-4414
This professional organization is dedicated to the improvement of all levels of education through the use of computer-based technology. Ask for a copy of ISTE's 40-page publications catalog offering special journals, books on computers in education, and educational software for educators and the classroom.

LEARNING SERVICES
P. O. Box 10636, Eugene, OR 97440-2636, 800-877-9376 West or Chelmsford, MA, 800-877-3278 East
Ask for 95-page catalog. Distributor of educational software from Broderbund, Davidson, The Learning Company, Software Toolworks, Roger Wagner, Optimum Resource, Microsoft, and Visions.

MEI/MICRO CENTER
1100 Steelwood Rd, Columbus, OH 43212   800-634-3478
Ask for 40-page catalog. This mail order retail distributor has competitive prices on just supplies for computers, printers, multimedia needs, 8mm video, and FAX machines.

MIDWEST VISUAL
6500 N Hamlin, Chicago, IL 60645   312-478-1250
This distributor of audiovisual and computer supplies sells software at an educator discount.

SCHOLASTIC SOFTWARE
Scholastic, Inc., P. O. Box 7502, 2391 E McCarty St, Jefferson City, MO 65102   800-541-5513
Ask for 66-page catalog offering software, CD-ROM's, videodiscs, and network applications, including 10 pages on science education materials.

SOFTWARE PLUS
Academic Inc., 50 E Palisade Ave, # 200, Englewood, NJ 07631   Catalog/Info 201-569-6262   Order line 800-377-9943
Ask for 24-page catalog from this software distributor. Significant discounts to students, teachers, and schools on latest versions of major software and CD-ROM's.

THE SOFTWARE SOURCE CO., INC.
2517 Hwy 35, Bldg N, Suite 201, Manasquan, NJ 08736   800-289-3275
Ask for 20-page catalog of software at large discounts for educators.

SUNBURST
101 Castleton St, P. O. Box 100, Pleasantville, NY 10570-0100   800-321-7511
Ask for 88-page catalog of multimedia software programs for elementary, middle, and high school students in math, science and computing.

TERRAPIN SOFTWARE, INC.
400 Riverside St, Portland, ME 04103-1068   800-972-8200
Ask for 16-page catalog of software programs for elementary school students in math, science and computing.

VERNIER SOFTWARE
2920 SW 89th St, Portland, OR 97225   503-297-1760
Software and hardware for the chemistry and physics laboratory. Ask for 30-page catalog.

Chapter 03

Education

Science Education Sources

These specialized information centers dedicated to science education can be of invaluable assistance.

ILLINOIS STATE BOARD OF EDUCATION
Contact Guen Pollack, N-242, Illinois State Board of Education, 100 N First, Springfield, IL 62777 217-782-2826
Ms. Guen Pollack is the the Senior Science Consultant for the Illinois State Board of Education.

MIDWEST CONSORTIUM FOR MATHEMATICS AND SCIENCE EDUCATION - NATIONAL NETWORK OF EISENHOWER MATHEMATICS AND SCIENCE
Contact Barbara Sandall, Dissemination Coordinator North Central Regional Educational Laboratory (NCREL), 1900 Spring Rd, Suite 300, Oak Brook, IL 60521-1480 708-218-1268 69 pages
Funded by the U. S. Department of Education, this consortium works to make connections between research and practice, among schools, and between schools

Chapter 03 - Education   41

and communities. Whether you are a teacher or a parent, this consortium is available to serve you by providing information, technical assistance, conferences, and products.

NATIONAL DIFFUSION NETWORK
Contact Dr. Shirley Menendez, Project Director, Illinois Statewide Facilitator Center, 1105 E Fifth St, Metropolis, IL 62960 618-524-2664
Illinois facilitator of the National Diffusion Network that produces reference works for educators, like *Mathematics, Science, and Technology Programs that Work*, a collection of 64 exemplary education programs.

NATIONAL SCIENCE EDUCATION STANDARDS
2101 Constitution Ave, NW, HA 486, Washington, DC 20418   202-334-1399
Ask to be placed on mailing list. National Committee on Science Education Standards and Assessment, National Research Council.

NATIONAL SCIENCE RESOURCES CENTER (NSRC)
Smithsonian Institution - National Academy of Sciences, Smithsonian Institution, Arts & Industries Building, Room 1201, Washington, DC 20560   202-357-2555
The NSRC works to improve the teaching of science in the nation's schools. NSRC disseminates information about effective science teaching resources, develops curriculum materials, and sponsors outreach and leadership development activities. Ask to be placed on the mailing list for the NSRC Newsletter.

PROMISING PRACTICES IN MATHEMATICS & SCIENCE EDUCATION
North Central Regional Educational Laboratory (NCREL), 1900 Spring Rd, Suite 300, Oak Brook, IL 60521-1480   708-571-4700   800-356-2735   158 pages
This book describes a collection of promising educational programs and practices used in schools across the United States.

SCIENCE & TECHNOLOGY INFORMATION SYSTEM (STIS)
National Science Foundation, 4201 Wilson Blvd, Washington, DC 22230   Voice 703-306-0214, Modem 7E1F 703-306-0212 or 703-306-0213
STIS is an electronic dissemination system that provides fast, easy access to NSF publications. There is no cost to you except for possible long-distance phone charges. The service is available 24 hours a day. Science education programs receive funding through the Division of Undergraduate Education, Directorate for Education and Human Resources, National Science Foundation.

Schools and Academies K-12

Various K-12 schools focusing on science in or near Chicago are included in the following list. Contact your local school board for information about K-12 science schools in suburban locations.

AMUDENSEN HIGH SCHOOL
5110 N Damen Ave, Chicago, IL 60625   312-534-2320   Pubic High School
This magnet high school has an environmental education program.

BARNARD COMPUTER, MATHEMATICS & SCIENCE CENTER
10354 S Charles, Chicago, IL 60643   312-535-2625   Public Elementary School
A science school of the Chicago Public Schools.

BOGAN COMPUTER TECHNICAL HIGH SCHOOL
3939 W 79th St, Chicago, IL 60652   312-535-2180   Public High School
A science school of the Chicago Public Schools.

GALILEO SCHOLASTIC ACADEMY OF MATH AND SCIENCE
820 S Carpenter, Chicago, IL 60607   312-534-7070   Public Elementary School
A science school of the Chicago Public Schools.

GERSHWIN MATH AND SCIENCE COMMUNITY ACADEMY
6206 S Racine, Chicago, IL 60636   312-535-9250   Public Elementary School
A science school of the Chicago Public Schools.

ILLINOIS MATHEMATICS AND SCIENCE ACADEMY
1500 W Sullivan Rd, Aurora, IL 60506-1000   708-907-5027   Public High School
IMSA is a high school academy in Illinois that offers rigorous courses in mathematics, science, art and humanities while emphasizing interconnections between disciplines. Students live in a pioneering educational community. Illinois residents who have completed nine years of education and are not enrolled in school beyond the ninth grade are invited to apply to the Academy. For further information contact the Academy.

LANE TECHNICAL HIGH SCHOOL
2501 W Addison St, Chicago, IL 60618   312-534-5400   Public High School
This magnet high school has a strong science program.

NEWBERRY MATHEMATICS AND SCIENCE ACADEMY
700 W Willow St, Chicago, IL 60614   312-534-8000   Public Elementary School
A science school of the Chicago Public Schools.

Chapter 03 - Education 43

RYDER MATH AND SCIENCE SCHOOL
8716 S Wallace, Chicago, IL 60620 312-535-3843 Public Elementary School
A science school of the Chicago Public Schools.

SCIENCE & ARTS ACADEMY
1825 Miner St, Des Plaines, IL 60016 708-827-7880 Private Elementary School
The Academy admits gifted students ages 3 1/2 through 15 years. Its curriculum is thematic and interdisciplinary.

For the first time these students are dissecting a frog in their science class. Life science becomes interesting to study when one puts aside common fears.

SHERIDAN MATH AND SCIENCE ACADEMY
533 W 27th, Chicago, IL 60616 312-534-9120 Public Elementary School
A science school of the Chicago Public Schools.

SUMNER MATH AND SCIENCE ACADEMY
4320 W Fifth Ave, Chicago, IL 60624 312-534-6730 Public Elementary School
A science school of the Chicago Public Schools.

44 Science Fun in Chicagoland

VON STEUBEN METROPOLITAN SCIENCE CENTER
5039 N Kimball Ave, Chicago, IL 60625   312-534-5100   Public High School
A science school of the Chicago Public Schools.

WHITNEY YOUNG HIGH SCHOOL
211 S Laflin, Chicago, IL 60607 312-534-7500   Public High School
This magnet high school has a strong science program.

Science Education Opportunities Science

Opportunities to learn about science and science teaching are listed below. These special opportunities are for children, adults, and teachers.

ACCESS 2000
Contact Eric Hamilton, Ph.D., Director   ACCESS 2000, Loyola University of Chicago, 6525 N Sheridan, Chicago, IL 60626   312-508-3582
ACCESS 2000 is a partnership of institutions in Chicago, including universities, academies and science laboratories, that participate in programs promoting full access to mathematics, science and engineering by the year 2000 for precollege students. Contact the ACCESS 2000 office for current program opportunities in over forty different programs in the Chicago area!

ADLER PLANETARIUM - DEPARTMENT OF EDUCATION
1300 S Lake Shore Drive, Chicago, IL 60605   312-322-0323
Contact the Department of Education about current programs and workshops available to the public.

AMERICAN RADIO RELAY LEAGUE
- DEPARTMENT OF EDUCATIONAL ACTIVITIES
American Radio Relay League, 225 Main St, Newington, CT 06111   203-666-1541
This is the best resource of information about Amateur Radio for all ages. Inquire about educational materials available including video courses, Morse Code and Ham Radio materials, Amateur Radio in the Classroom newsletter, license publications for students, and reference books. Ask for list of School Amateur Radio Clubs and Advisors.

ARGONNE NATIONAL LABORATORY
- DIVISION OF EDUCATIONAL PROGRAMS
Division of Educational Programs DEP/223, Argonne National Laboratory, 9700 S Cass Ave, Argonne, IL 60439-4845   708-252-4579
Twenty different programs at Argonne are available in research and education for college/university faculty and students. For information about available programs

telephone 708-252-4579. For information about the Regional Instrumentation Sharing Program telephone 708-252-9818. For information about workshops, institutes and conferences telephone 708-252-2573.

ARGONNE NATIONAL LABORATORY
- ENVIRONMENTAL EDUCATION PROGRAMS
Division of Educational Programs DEP/223, Argonne National Laboratory, 9700 S Cass Ave, Argonne, IL 60439-4845   708-252-7613
Educational programs in environmental education are available for grades 5-12. Also inquire about *An Environmental Education Resource List* available to teachers grades K-12+ that contains over 200 references to educational resources in the Chicago area and nationwide.

BROOKFIELD ZOO
- EDUCATION DEPARTMENT SCHOOL RESERVATIONS
3300 S Golf Rd, Brookfield, IL 60513   708-485-0263, ext 367
Ask for school brochure describing field trips for school groups, educator workshops, and programs to enhance classroom studies.

CASPAR
- CHICAGO AREA SCIENCE PROGRAM APPLICATION RESOURCE
ACCESS 2000, Loyola University of Chicago, 6525 N Sheridan, Chicago, IL 60626   312-508-3582
This free service to Chicago area students and their parents, teachers and school administrators provides information on opportunities for students to participate in summer science programs in the Chicago area. Contact the ACCESS 2000 office above for current information.

CHANNEL 2 TELEVISION NEWS WEATHER TEAM
Paul Douglas, Steve Baskerville, and Harry Volkman   Channel 2, Chicago, IL 312-951-3631
If you would like information about how a member of the weather team can visit your child's school, call the Channel 2 Newsroom at 312-951-3631. Learn about weather topics like Doppler 2000 radar, Earthwatch, Stormwatch, Watches & Warnings, Tornado Warning, and Severe Weather Conditions.

CHICAGO ACADEMY OF SCIENCES - EDUCATION DEPARTMENT
The Chicago Academy of Sciences, 2001 N Clark St, Chicago, IL 60614   312-549-0606
Contact the Education Department about current programs and classes for both teachers and children, like Science in the Park, Nature Exploration Labs, Water Works Lab, and Discovery Tours. Special events include HerPETological Weekend, held annually in September. Ask for a copy of *Nature's Notes*, a newsletter of events for friends of the Academy.

CHICAGO BOTANIC GARDEN - REGISTRAR'S OFFICE
Chicago Botanic Garden, P. O. Box 400, 1000 Lake-Cook Road (at Edens Expressway), Glencoe, IL 60022-0400   708-835-8261
Call the registrar's office to obtain a Course Guide listing many courses including Bird Walk, Preschool Story Time, Prairies Then and Now, Floral Design, Continuing Gonsal, Bulbs for Spring Gardens, and many others. Lower registration fees for Garden members.

CHICAGO BOTANIC GARDEN
- SCHOOL PROGRAMS FOR CHILDREN AND TEACHERS
Contact Anne Grall Reichel, Supervisor of Teacher Services at 708-835-8323 and Katherine Johnson, Coordinator, School Programs at 708-835-8279. Chicago Botanic Garden, P. O. Box 400, 1000 Lake-Cook Road (at Edens Expressway), Glencoe, IL 60022-0400
Inquire about hands-on teacher workshops for graduate credit, including Workshops for Students, the New Explorers Program, the Environmental Education Awareness Program, and the Collaborative Outreach Education Program that brings the wonders of plant science to inner city students and teachers.

CHICAGO CHILDREN'S MUSEUM
- CITY STALKERS SUMMER CAMP
North Pier Chicago, 435 E Illinois St, (Moving to 600 E Grand Ave, May, 1995), Chicago, IL 60611   312-527-1000
At one of the summer camp sessions children learn simple things you can do to preserve the environment and what the city is doing to conserve our resources.

CHICAGO CHILDREN'S MUSEUM - TRAVELING TRUNK SHOWS
North Pier Chicago, 435 E Illinois St, (Moving to 600 E Grand Ave, May 1995), Chicago, IL 60611   312-527-1000
The museum staff will take to the road and bring exhibit materials to your classroom, including Buildings and Lego, The Stinking Truth about Garbage, and Smarter than You Think. Teachers can also arrange for class field trips to the museum.

CHICAGO SYSTEMIC INITIATIVE (CSI)
Contact Adrian D. Beverly, Assistant Superintendent, Department of Instructional Support. Chicago Public Schools, 1819 W Pershing Rd, Chicago, IL 60609   312-535-8850
Major National Science Foundation (NSF) funded initiative to promote system-wide reform and improvement in mathematics, science and techology instruction. Limited to the Chicago Public Schools. Initial participation is negotiated through district offices of the Chicago Public Schools.

Chapter 03 - Education 47

COLUMBIA COLLEGE CHICAGO
- INSTITUTE FOR SCIENCE EDUCATION
Contact Zafra Lerman, Director, Institute for Science Education, Columbia College Chicago, 600 S Michigan Ave, Chicago, IL 60605   312-663-1600, ext 180
Contact the Institute about current programs and workshops. The Institute is dedicated to providing informative and motivating introductory science for today's students, especially students of the arts.

DISCOVERY CENTER MUSEUM
711 N Main St, Rockford, IL 61103 815-963-6769
The Museum has over 100 hands-on science exhibits inside the museum and even more in the outdoor science park.   Ask for the educator's guide to Discovery Center listing traveling exhibits, special events, and school field trip information.

FIELD MUSEUM OF NATURAL HISTORY
- ADULT AND FAMILY PROGRAMS
The Field Museum, Roosevelt Road at Lake Shore Drive, Chicago, IL 60605-2497 312-922-9410, ext 854   Fees vary from $ 7 to $ 70.
Contact the following divisions for current activities brochure: Natural History Field Trips 312-922-9410, extension 362; Children's Workshops 312-922-9410, extension 399; and Adult Courses, Performances, Lectures 312-922-9410, extension 575.

FIELD MUSEUM OF NATURAL HISTORY - RESOURCE CENTERS
The Field Museum, Roosevelt Road at Lake Shore Drive, Chicago, IL 60605-2497   312-922-9410
The Africa Resource Center has books and audiovisuals, telephone 312-922-9410, extension 883; The Webber Resource Center on Indians of the Americas has books, audiovisuals, and hands-on materials, telephone 312-922-9410, extension 497; and the Rice Wildlife Research Station has books, materials and computers on wildlife, telephone 312-922-9410, extension 814. Teachers should telephone ahead for class visits or for current hours to use materials.

FIELD MUSEUM OF NATURAL HISTORY
- SCHOOL AND COMMUNITY SERVICES
The Field Museum, Roosevelt Road at Lake Shore Drive, Chicago, IL 60605-2497 312-922-9410, ext 852   Illinois school and community groups free admission.
Contact these divisions for current program brochure: School tours & registration information 312-922-9410, extension 353; Community Outreach 312-922-9410, extension 363; and Teacher Services 312-922-9410, extention 365.   Teacher training fees vary from $ 10 to $ 223 for graduate credit.

48 Science Fun in Chicagoland

FOREST PRESERVE DISTRICT OF COOK COUNTY
Conservation Department, 536 N Harlem Ave, River Forest, IL 60305 708-771-1330
Ask for brochure, *Getting in Touch with Nature, Classroom Programs for Cook County Schools*, including classroom programs, slide programs, and visiting nature centers. Nature centers include Crabtree Nature Center, Barrington, IL 708-381-6592; River Trail Nature Center, Northbrook, IL 708-824-8360; Trailside Museum, River Forest, IL 708-366-6530; Little Red Schoolhouse Nature Center, Willow Springs, IL 708-839-6897; Sand Ridge Nature Center, South Holland, IL 708-868-0606; and Camp Sauawau, Lemont, IL 708-257-2045. Please schedule your program as early as possible, at least three weeks in advance.

FOREST PRESERVE DISTRICT OF DUPAGE COUNTY
P. O. Box 2339, Glen Ellyn, IL 60138   708-790-4900
Ask for 40 page booklet, *Let's Have a Class Outside Today: A Teacher's Guide to the Forest Preserve District of Dupage County*. Nature centers include Fullersburg Woods Environmental Education Center, Oak Brook, IL; Willowbrook Wildlife Center, Glen Ellyn, IL; and Kline Creek Farm, West Chicago, IL.

GOLDEN APPLE SCIENCE PROGRAM
Contact Jane A. Rosen, Ph.D., Executive Director, Golden Apple Foundation, 8 South Michigan Ave, Suite 700, Chicago, IL 60603-3318   312-407-0006
Summer workshops for elementary school teachers in city, suburban, public, parochial, and independent schools that give teachers an opportunity to develop hands-on science programs. Teachers receive stipends and grants. This program allows teacher participants to network with and receive ongoing support from colleagues and expert teachers.

SCHOOL OF HOLOGRAPHY
Contact Dr. Ted Niemiec, Director of Education School of Holography, 1134 W Washington Blvd, Chicago, IL 60607   312-226-1007   Private School
The comprehensive curriculum at the School of Holography allows the student to explore holography as an artist, as a scientist, and as an engineer. Twelve different courses are offered from introductory to advanced instruction. Tuition fees vary from $ 175 to $ 375 with laboratory fees and materials extra.

ILLINOIS DEPARTMENT OF ENERGY AND NATURAL RESOURCES
325 W Adams St, Room 300, Springfield, IL 62704-1892   800-252-8955
Inquire about the Illinois Energy Education Development (ILEED) Program and ask about newsletters, free materials, and workshops and in-school programs for students and teachers. Ask for brochure of free publications, *Educational Materials Available from the ENR Information Clearinghouse*.

Chapter 03 - Education 49

INSTITUTE FOR MATHEMATICS AND SCIENCE EDUCATION
Contact Howard Goldberg and Philip Wagreich, Co-Directors, Institute for Mathematics and Science Education, Room 2075 Science & Mathematics Laboratories, University of Illinois at Chicago, 840 W Taylor St, Chicago, IL 60607   312-996-2448
Contact the Institute about current programs, workshops and materials. Ask for the *Documents Catalog* that describes ordering information about TIMS (Teaching Integrated Mathematics and Science) curriculum and laboratory materials. TIMS, in use since 1974, is a hands-on quantitative approach to K-8 science that uses fun experimental methods and thinking to integrate math and science.

INVENTOR'S COUNCIL
Contact Don Moyer, President Inventor's Council, 431 S Dearborn 705, Chicago, IL 60605 312-939-3329
The Inventor's Council holds inventor's workshops on one Saturday morning each month at the Harold Washington Library Center, 400 S State St, Chicago, Illinois. Topics include: How to Evaluate Patents, Patent Sucess Stories, How to Get the Best Patent, Finding the Best Way to Get New Products Made, How to Get Manufacturers to Invest in Inventions. This not-for-profit Council asks for contributions and supplies write-ups on various topics.

JOHN G. SHEDD AQUARIUM - PUBLIC PROGRAMS
Contact Bert Vescolani, Education Department, John G. Shedd Aquarium, 1200 S Lake Shore Drive, Chicago, IL 60605   312-939-2426, ext 3359   Fees range from $ 12 to $ 240 Age: Preschool through 12th Grade, Adult
Special courses are available to the public in classrooms and laboratories at the John G. Shedd Aquarium. Scheduled at various times these classes cover a multitude of fun topics from pond monsters to killer plants.

JOHN G. SHEDD AQUARIUM - SCHOOL PROGRAMS
Contact: Lynne Hubert, Education Department, John G. Shedd Aquarium, 1200 S Lake Shore Drive, Chicago, IL 60605   312-939-2426 Various times Monday through Friday   Free to Illinois schools ($ 10 lab fee)   Age: Kindergarten through college.
Special classes and labs from 30 minutes to 2 hours are available to school groups at the John G. Shedd Aquarium. Call 312-986-2300 for reservations and the *School Brochure*. Teachers' workshops are offered thoughout the year.

JURICA NATURE MUSEUM
Contact Fr. Theodore Suchy, Scholl Science Center, Illinois Benedictine College, 5700 College Rd, Lisle, IL 60532 708-960-1500, ext 1515
Loan program available to teachers. Visitors experience the African savanna, tropical rain forest, the woodlands, wetlands and praire of Northern Illinois along with other smaller animal habitat dioramas. Insects, fish, reptiles and all kinds of birds are displayed along with many fossil specimens and skeletons.

50 Science Fun in Chicagoland

These two students are obviously enjoying the Junior Engineering Technical Society (JETS) annual engineering design competition.

LAKE COUNTY EDUCATIONAL SERVICE CENTER
Contact Fredric Tarnow, Science Coordinator, Lake County Educational Service Center, 19525 W Washington, Grayslake, IL 60030   708-223-3400
Ask for calendar of workshops listing several mathematics and science workshops offered throughout the school year. Fees vary and workshops are open to all schools unless specified for COOP MEMBERS only.

LAKE COUNTY FOREST PRESERVES
2000 N Milwaukee Ave, Libertyville, IL 60048-1199   708-948-7750
Ask for *Group Environmental Education Program Guide* with information about educational programs, prices, reservations, locations, dates, self-guided programs, and naturalist-guided programs. Over 20 programs available.

Chapter 03 - Education

LEON M. LEDERMAN SCIENCE EDUCATION CENTER - MATH SCIENCE CONSORTIA DEMONSTRATION SITE

Fermi National Accelerator Laboratory, P. O. Box 500 MS 777, Batavia, IL 60510   708-840-8258

At this science education center the visitor may preview the Internet, science education computer bulletin boards, and see multimedia materials. The Teacher Resource Center is a demonstration site for mathematics and science materials in the Math Science Consortia, North Central Regional Educational Laboratory for the Eisenhower National Clearinghouse. Phone for an appointment.

LEON M. LEDERMAN SCIENCE EDUCATION CENTER - PRECOLLEGE EDUCATION PROGRAMS

Fermi National Accelerator Laboratory, P. O. Box 500 MS 777, Batavia, IL 60510   708-840-8258

Contact this center for current information about programs for high school teachers and students, programs for elementary and midlevel schools, programs for students of all ages, and science materials programs ---in total over 50 different programs. This Science Education Center is equipped with a technology classroom, classroom laboratory, and hands-on exhibits dedicated to physical science concepts related to Fermilab. Science toys are available in a small retail shop.

LEON M. LEDERMAN SCIENCE EDUCATION CENTER - TEACHER RESOURCE CENTER

Contact Susan Dahl, Coordinator, Teacher Resource Center, Fermi National Accelerator Laboratory, Leon M. Lederman Science Education Center, P. O. Box 500 MS 777, Batavia, IL 60510   708-840-8258   Monday-Friday 8:30-5:00; Saturday 9:00-3:30. Call for an appointment.

This extensive teacher resource center is filled with books, periodicals, kits, videotapes, etc. - for teachers, administrators, librarians, scientists, and Science Center program participants.

THE MORTON ARBORETUM

Route 53 (just north of interstate 88), Lisle, IL 60532   708-719-2400

Ask for a copy of the quarterly, *The Morton Arboretum*, a newletter listing events, news and classes available. Also ask for a copy of the 25 page resource booklet, *Selecting & Planting Trees*.

MOTOROLA MUSEUM OF ELECTRONICS - EDUCATIONAL PROGRAMS

Contact the Manager of Programs and Services, Motorola Museum of Electrons, Motorola, Inc., 1297 E Algonquin Rd, Schaumburg, IL 60196   708-576-6400

This museum provides interactive exhibits which chronicle the revolution in electronics technology over the 20th Century. Contact the Manager of Programs and Services about current educational programs available for students grades 6-12.

52 Science Fun in Chicagoland

MUSEUM OF SCIENCE & INDUSTRY - EDUCATION DEPARTMENT
Museum of Science & Industry, 57th St and Lake Shore Drive, Chicago, IL 60637 312-684-1414, ext 2429

Inquire about workshops for teachers and staff members in staff development by contacting Ed McDonald at ext 2423. Live interpretation and demonstrations of a variety of science and technology principles are offered within the Museum exhibits; contact Diann Milford at ext 2463. Contact Kirsten Ellenbogen at ext 2402 for activity guides that support Museum exhibits and offer teachers ideas for classroom activities before and after their class visit. For information about science fairs held annually at the Museum contact Ed McDonald at ext 2423.

MUSEUM OF SCIENCE & INDUSTRY - NASA TEACHER RESOURCE CENTER
Contact Ed McDonald, Museum of Science & Industry, 57th St and Lake Shore Drive, Chicago, IL 60637 312-684-1414, ext 2423

This teacher resource center is open by appointment. It contains books, slides, video tapes, posters and materials about the United States space program.

MUSEUM OF SCIENCE & INDUSTRY - NEW EXPLORERS
Contact Sue Harrison Museum of Science & Industry, 57th St and Lake Shore Drive, Chicago, IL 60637 312-684-1414, ext 2480

The popular PBS television series, *The New Explorers*, is made available to teachers along with educational materials. Workshops and tape support groups also are provided for teachers.

ORIENTAL INSTITUTE MUSEUM - EDUCATION DEPARTMENT
Oriental Institute Museum, University of Chicago, 1155 E 58th St, Chicago, IL 60637 312-702-9507

The Museum exhibits the cultural history of the ancient Middle East. Contact the Education Department about current programs.

THE POWER HOUSE - EDUCATIONAL PROGRAMS
Commonwealth Edison, 100 Shiloh Blvd, Zion, IL 60099 708-746-7492

Teachers and schools can request the following educational programs for presentation to students at The Power House: Alternative Energy Sources, The Nature of Energy, Energy and the Environment, Energy Conservation, Wonders of Electricity, and Power Generation, all with pre- and post-visit materials. (A program that visits schools, Safety and Electricity, is available for K-3.) Ask for brochure on tours and educational programs.

THE POWER HOUSE - ENERGY RESOURCE CENTER
Contact Mary L. Crompton, Energy Education Assistant   Commonwealth Edison, 100 Shiloh Blvd, Zion, IL 60099 708-746-7850   Hours: Tuesday-Saturday 10:00-5:00

This science education resource center is open to both students and teachers for research and study. It contains over 500 books, 48 periodicals, six computers and four VCR's with video monitors.

SCIENCE 2001 TEXT SETS
Contact Dr. Diane Schiller, Loyola University Chicago, 6525 N Sheridan Rd, Chicago, IL 60626   312-508-8383
Teachers, grades K-12, can borrow sets of science books grouped by subject, like inventors, insects, animals, space, weather, water, earth science, states of matter, the egg, maple sugaring, human body, rainforests, science of music and spiders.

SCIENCE DISCOVERY CENTER
H. R. McCall School, 3215 N McAree, Waukegan, IL 60087   708-360-5480
A science resource center for use by local district schools is now available for student fun. Built from a portable classroom this center is equipped with work stations providing hands-on science from real dinosaur eggs to giant swinging bowling ball pendulums. School systems interested in developing a science resource center should contact the school principal.

THE SCIENCE EXPLORERS PROGRAM
Write LaVonia M. Ousley, Program Coordinator, Division of Educational Programs, Argonne National Laboratory DEP/223, 9700 S Cass Ave, Argonne, IL 60439-4845   708-252-7784
Request *The Science Explorers Program Information Packet* by writing the Program Coordinator at the above address. This packet includes a list of regional facilities and educational programs.

SCIENCE LINKAGES IN THE COMMUNITY (SLIC) - AMERICAN ASSOCIATION FOR THE ADVANCEMENT OF SCIENCE
Contact Michael Hyatt, Community Developer AAAS-SLIC, Granada Centre #220, 6525 N Sheridan, Chicago, IL 60626   312-508-8756
This new program of the AAAS was created to organize science opportunities for students in the urban areas and within community organizations. Contact the SLIC office about current opportunities in your area.

SCIENCE-BY-MAIL
Museum of Science, Science Park, Boston, MA 02114-1099   800-729-3300
$ 56 per membership group of one to four kids.
Kids grades 4-9 complete science packets with fun investigations and mail them to a pen-pal scientist for encouraging feedback.

SCITECH - THE SCIENCE AND TECHNOLOGY INTERACTIVE CENTER - EDUCATION PROGRAMS
Education Programs Coordinator, SciTech, 18 W Benton, Aurora, IL 60506   708-859-3434
Programs and workshops available at SciTech include SciTech Clubs for Girls,

54 Science Fun in Chicagoland

lectures and demonstrations, Summer Science Camp, early childhood programs, and Overnights. Fees vary from $ 15 to $ 120. Contact the Education Programs Coordinator for current information.

SPRING VALLEY ENVIRONMENTAL EDUCATION OUTREACH PROGRAM
Spring Valley Nature Sanctuary, 1111 E Schaumburg Rd, Schaumburg, IL 60194   708-980-2100
Teachers, invite a naturalist from Spring Valley Nature Sanctuary to visit your class for a one hour lesson including games and activities.

TEACHERS ACADEMY FOR MATHEMATICS AND SCIENCE (TAMS)
10 W 35th St, Chicago, IL 60616   (On the campus of Illinois Institute of Technology) 312-808-0100
Founded by Leon Lederman, Nobel laureate in physics, TAMS provides many opportunities for teacher training in mathematics and science. Over 40 schools chosen and interested in staff development now receive ongoing teacher training and educational materials assistance. Technology and science workshops for individual teachers in the Chicago Public Schools are also available through the Academy. TAMS also sponsors conferences on mathematics and science education. Contact Sherry Bushre about technology training; Yolanda Nellum about mathematics and science training; and Helen Chang about staff development services.

VENTURES IN SCIENCE
Contact Elena Mulcahy, Ed.D., Director, Ventures in Science, Truman College, 1145 W Wilson Ave, Chicago, IL 60640   312-907-4097
The purpose of Ventures in Science is to motivate and prepare high school students from minority groups who are underrepresented in mathematics and science professions. For information about current activities and programs contact the director.

WALTER E. HELLER NATURE CENTER
2821 Ridge Rd, Highland Park, IL 60035   708-433-6901   Hours: Monday-Saturday 8:30-5:00; Sunday 10:00-4:00. Park hours: 6:00-9:00 seven days a week. Admission free.
This Center is a 97-acre forest with more than three miles of marked trails, including a building with a community room, a classroom, and a reference library. Request a *Park District of Highland Park Catalog* listing 80 pages of special events, educational nature classes, programs for school groups preschool through high school, and athletic programs.

Chapter 04

Events

Local Events, Competitions, and Awards

AMERICAN CHEMICAL SOCIETY
- CHICAGO SECTION ANNUAL SCHOLARSHIP EXAMINATION
American Chemical Society, Chicago Section, 7173 N Austin, Niles, IL 60714 **708-647-8405**
Each year in May the Chicago Section of the American Chemical Society sponsors an Annual Scholarship Examination for Chicagoland high school students. This exam has been administered annually on the campus of the University of Illinois at Chicago.

ARGONNE NATIONAL LABORATORY
Office of Public Affairs, Argonne National Laboratory, 9700 S Cass Ave, Argonne, IL 60439-4845 **708-252-5562**
For special events open to the public contact Argonne's Office of Public Affairs for current information.

56 Science Fun in Chicagoland

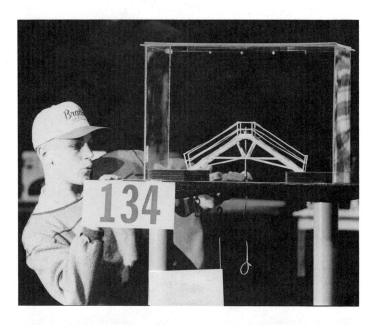

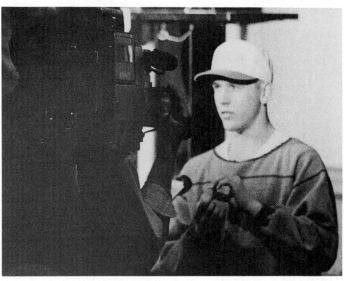

This winning bridge design at the IIT Bridge Building Contest resulted in an evening news television interview for this young student engineer.

THE ANNUAL BRIDGE BUILDING CONTEST
Contact Department of Physics, Siegel Hall, Illinois Institute of Physics, Chicago, IL 60616 312-567-3375
This engineering design contest is annually held in February on the campus of IIT. Ask for rules and information about availability of bridge building kits. This popular event sends the top two winners to the International Bridge Building Contest and provides awards and scholarships to winning bridge designs.

BROOKFIELD ZOO - SPECIAL EVENTS
3300 S Golf Rd, Brookfield, IL 60513   708-485-0263, ext 365
Ask for brochure describing the many special events held each year at Brookfield Zoo.

CHICAGO CHILDREN'S MUSEUM - ENVIROMANIA
North Pier Chicago, 435 E Illinois St, (Moving to 600 E Grand Ave, May, 1995), Chicago, IL 60611 312-527-1000   Annual event
At this special event children learn about their environment.

CHICAGO CHILDREN'S MUSEUM - INVENTING FAIR
North Pier Chicago, 435 E Illinois St, (Moving to 600 E Grand Ave, May, 1995), Chicago, IL 60611 312-527-1000   Annual event
Contact the museum for information about this event.

CHICAGO CHILDREN'S MUSEUM - RECYCLED RALLY
North Pier Chicago, 435 E Illinois St, (Moving to 600 E Grand Ave, May, 1995), Chicago, IL 60611 312-527-1000   Annual event
At this special event children will learn about non toxic cleaning solutions, recycle bins, and compost. Children bring unwanted materials (trash) that can be sold to others at this event to learn about salvage.

CHICAGOLAND SKY LINERS KITE CLUB
Contact Tom and Leora McCune Chicagoland Sky Liners Kite Club, 981 Twisted Oak, Buffalo Grove, IL 60089 708-537-7066
The 11th annual Sky Circus was held at Schaumburg, Illinois, in 1994. For current information about kite festivals in the U. S. and internationally see the current *KiteLines* magazine for *Pocket Kite Calendar and Almanac*.

FIELD MUSEUM OF NATURAL HISTORY - VISITOR PROGRAMS
The Field Museum, Roosevelt Road at Lake Shore Drive, Chicago, IL 60605-2497 312-922-9410, ext 658
Contact these divisions for current activities brochure and information: Webber Resource Center for Indians/Americas 312-922-9410, extension 497; Weekend Programs and Activities 312-922-9410, extension 288 or 350.

GOLDEN APPLE FOUNDATION
Contact Jane A. Rosen, Ph.D., Executive Director, 8 South Michigan Ave, Suite 700, Chicago, IL 60603-3318   312-407-0006
Nonprofit organization that recognizes excellent teachers K-12 in Cook, Lake, and DuPage Counties. Nomination period is September through November each year. Golden Apple Award-winners become part of the foundation's Academy of educators, creating programs to recruit and renew teachers.

THE IIT 100 SPEEDWAY
Illinois Institute of Technology, Alumni Office - IIT 100, 10 W 35th St, Chicago, IL 60616   312-808-5466
The IIT 100 race is annually held in April. Students compete with student engineered cars for monetary prizes and a pizza party. Ask for rules and information about this popular event. Divisions include high school, college, and alumni/faculty. Cost of registration is $ 25.00 per team, which will provide a kit to construct the vehicle. This event is sponsored by Snap-On Tools and Ford Motor, Inc.

IIT INDUSTRIAL DESIGN OPEN HOUSE
School of Industrial Design, IIT, 35th & State St, Chicago, IL 60616   312-808-6500   Annually in May
This one day event allows guests to meet designers and student designers for brief consultations about their invention designs.

THE ILLINOIS SCIENCE OLYMPIAD
c/o Jeremy Way, 505 S Mathews Ave, Box 57-1, Urbana, IL 61801-6317   217-337-6582
Ask to be placed on the mailing list to receive a newsletter. The goal of this organization is to improve the quality of science education and is accomplished through classroom activities and the encouragement of tournaments where student teamwork accomplishes scientific tasks in all areas of applied science, including biology, chemistry, physics, earth sciences and engineering. Divisions include A1 (grades K-3), A2 (grades 3-6), B (grades 5-9), and C (grades 9-12).

JETS ENGINEERING DESIGN COMPETITION
Contact David L Powell, Director, Illinois JETS, University of Illinois at Urbana-Champaign, College of Engineering, 1304 W Green St, Room 207, Urbana, IL 61801-2982   800-843-5410
Annually Illinois JETS (Junior Engineering Technical Society) conducts written exam competition as well as engineering design competition in Chicago. Ask for current information and a copy of the *Engineering Library for Pre-College Students and Teachers Catalog* of brochures, books, and videotapes. Illinois JETS has offered programs since 1959.

The IIT 100 race is a fun, popular competition annually held each April for student engineered cars.

LOYOLA UNIVERSITY FAMILY SCIENCE NIGHT
Contact Dr. Diane Schiller, Loyola University Chicago, 6525 N Sheridan Rd, Chicago, IL 60626 312-508-8383 Annually in March
This fun event includes student science fair projects, teacher text sets, technology and science history. A fun event for everyone in the community.

MUSEUM OF SCIENCE & INDUSTRY - FESTIVALS IN SCIENCE
Contact Ruth Goehmann, Museum of Science & Industry, 57th St & Lake Shore Drive, Chicago, IL 60637 312-684-1414, ext 2225
Festivals in science are held at the Museum where special activities are orchestrated around a theme or topic such as Earth Day or Space Day.

MUSEUM OF SCIENCE & INDUSTRY - TEACHER OPEN HOUSE
Contact Jane Peterson, Museum of Science & Industry, 57th St & Lake Shore Drive, Chicago, IL 60637 312-684-1414, ext 2429
The Teacher Open House is usually held twice a year, offering hands-on activities and an opportunity for teachers to be guests at the Museum to learn about new exhibits and programs that can support educational teaching goals.

60 Science Fun in Chicagoland

THE POWER HOUSE - SCIENCE ON SATURDAY
Commonwealth Edison, 100 Shiloh Blvd, Zion, IL 60099   708-746-7080   One Saturday each month.
Special programs and guests are at The Power House one Saturday each month. Ask for *Science on Saturday* brochure describing upcoming events.

SIX FLAGS GREAT AMERICA PHYSICS DAYS
Contact Lisa Ignoffo, Special Events Representative, Six Flags Great America, P. O. Box 1776, 542 N Route 21, Gurnee, IL 60031   708-249-2133 ext. 6439
Each May, Six Flags Great America hosts Physics Days. This successful program brings high school students from all over Chicagoland to the amusement park for educational fun in a recreational atmosphere. On the rides students measure their acceleration, horsepower, and centripetal force as they become a moving part of science experiments.

WIERD SCIENCE KIDS
Contact Lee Marek, Instructor, Naperville North High School, 899 N Mill St, Naperville, IL 60563-8998   708-420-6443
The Wierd Science Kids make chemistry fun with wild, mad science demonstrations. This group of teachers often go on the road with their show. You may have seen Lee Marek squirt television's David Letterman with a fire extinguisher, cover him with shredded styrofoam and blind him with burning magnesium. The Wierd Science Kids often make presentations to local audiences.

National Events, Competitions, and Awards

THE DUPONT CHALLENGE
Science Essay Awards Program, General Learning Corporation, 60 Revere Drive, Northbrook, IL 60062-1563   708-205-3000
The DuPont Challenge is an annual national science essay contest, Junior Divison grades 7-9 and Senior Division grades 10-12. Annually nearly $ 8,000 in educational grants for students are awarded. Each essay must be between 700 and 1,000 words and conform to all rules. Inquire about the current rules and deadlines to the Science Essay Awards Program at the above address.

DURACELL/NSTA SCHOLARSHIP COMPETITION
National Science Teachers Association, 1840 Wilson Blvd, Arlington, VA 22201-3000   703-243-7100
This competition awards saving bonds to young inventors in grades 9-12. First place winner will receive a $ 20,000 savings bond. Inquire about rules and deadlines.

Chapter 04 - Events 61

INTERNATIONAL SCIENCE AND ENGINEERING FAIR
Science Service, Inc., 1719 N St, NW, Washington, DC 20036   202-785-2255
For over forty years Science Service has supervised the International Science and Engineering Fair. More than 800 students participate from over 20 different countries. Students in grades 9-12 are eligible and two student finalists are selected from each of the 415 regional science fairs.

NASA/NSTA SPACE SCIENCE STUDENT INVOLVEMENT PROGRAM (SSIP)
National Science Teachers Association, 1840 Wilson Blvd, Arlington, VA 22201-3000   703-243-7100
This competition encourages students to work in a team to create and design futuristic aircraft or spacecraft; investigate the effect of human activity on the Earth's ecosystem; propose experiments that could be performed at NASA facilities or on space flights; or design an expedition to Mars. Inquire about rules and deadlines.

PRESIDENTIAL AWARDS FOR EXCELLENCE IN SCIENCE AND MATHEMATICS TEACHING (PAESMT)
Elementary PAESMT, National Science Teachers Association, 3140 N Washington Blvd, Arlington, VA 22201   703-243-7100   Secondary PAESMT, National Science Teachers Association, 5112 Berwyn Rd, 3rd Floor, College Park, MD 20740   301-220-0875
This program was established by The White House and identifies outstanding teachers of science and mathematics, K-12, who will serve as models for their colleagues and who will form a leadership core to help advance the major reform movements in these disciplines. Contact the Illinois Science Teachers Association for local nomination forms and nomination deadlines.

TANDY TECHNOLOGY SCHOLARS
P. O. Box 32897, TCU Station, Fort Worth, TX 76129   817-924-4087
High schools thoughout the United States may nominate outstanding students in math, science or computer science or may nominate an outstanding teacher in these areas to receive scholarships and awards. Application deadline is Mid-October.

TAPESTRY/NSTA
1840 Wilson Blvd, Arlington, VA 22201-3000   703-243-7100
Toyota's Appreciation Program for Excellence to Science Teachers Reaching Youth (TAPESTRY). Competition is open to science teachers of grades 6-12. $ 400,000 in grants to teachers is available. Inquire about rules and deadlines.

62 Science Fun in Chicagoland

TOSHIBA NSTA EXPLORAVISION AWARDS PROGRAM
National Science Teachers Association, 1840 Wilson Blvd, Arlington, VA 22201
800-EXPLOR-9
The purpose of the competition is to encourage students to combine their imagination with the tools of science and technology to create and explore a vision of the future. Each contest entry is limited by rules and consists of a project description (ten pages or less), bibliography and ten storyboard frames. Student members of the four first place teams will each receive a $ 10,000 U. S. savings bond.

WESTINGHOUSE SCIENCE TALENT SEARCH
Science Service, Inc., 1719 N St, NW, Washington, DC 20036 202-785-2255
Since 1942, Science Service, Inc., has administered this nationwide competition that has included five future Nobel Prize winners. High school seniors complete independent research projects and submit written reports on their findings. Finalists enjoy traveling to Washington, DC, and meet the President of the United States.

Chapter 05

Excursions

Excursion Reference Books

AMUSEMENT PARK PHYSICS: A TEACHER'S GUIDE
by Nathan A. Unterman J. Weston Walch, Publisher, 321 Valley St, P. O. Box 658, Portland, ME 04104-0658 800-341-6094 159 pages $ 19.95
This guide provides tutorials, practice problems, and lab exercises appropriate for studying the motion of amusement park rides.

CHICAGO'S MUSEUMS
by Victor J. Danilov Chicago Review Press, Inc., 814 N Franklin St, Chicago, IL 60610 291 pages $ 11.95
More than 150 museums are listed including address, telephone, admission, hours, and a description of the museum.

64 Science Fun in Chicagoland

ENVIRONMENTAL EDUCATION RESOURCE LIST
Contact Anne Marie Smith, Environmental Education Programs, Division of Educational Programs DEP/223, Argonne National Laboratory, 9700 S Cass Ave, Argonne, IL 60439-4845 708-252-7613
This extensive guidebook, available to teachers grades K-12+, contains over 200 references to environmental education resources and excursions available in the Chicago area.

EXPLORING SCIENCE: A GUIDE TO CONTEMPORARY SCIENCE AND TECHNOLOGY MUSEUMS
Association of Science-Technology Centers, 1025 Vermont Ave, NW, Suite 500, Washington, DC 20005 202-783-7200 1980 72 pages In print
This is a complete, but older, directory to science and technology museums across the United States as well as affiliate international member museums.

HIKING & BIKING IN LAKE COUNTY, ILLINOIS
by Jim Hochgesang Roots and Wings Publications, P. O. Box 167, Lake Forest, IL 60045 128 pages $ 10.95
For direction to nature excursions this book describes 25 different nature trails and forest preserves.

TEACH THE MIND, TOUCH THE SPIRIT: A GUIDE TO FOCUSED FIELD TRIPS
The Field Museum, Education Department, Roosevelt Road at Lake Shore Drive, Chicago, IL 60605-2497 312-922-2497, ext 351 80 pages $ 10.00
This book describes museums as educational opportunities, structuring your field trip, The Field Museum opportunities, and a reference bibliography.

TEN-MINUTE FIELD TRIPS
by Helen Ross Russell Distributed by the National Science Teachers Association 800-722-NSTA 1991 176 pages $ 16.95
More than 200 short, close-to-home excursions in science for grades K-8 are described. Each excursion is categorized by science subject area and lists classroom activities with teacher preparation needs. Excursions are described for both rural and urban locations. Fun for both teachers and parents.

TREKS FOR TROOPS: TROOP REFERENCE FOR TRIPS
Contact Girl Scouts of Chicago, 55 E Jackson Blvd, Suite 1400, Chicago, IL 60604 1994 38 pages
This reference lists over 100 places in Metropolitan Chicago where one can take a group.

Local Excursion Opportunities

ADLER PLANETARIUM
1300 S Lake Shore Drive, Chicago, IL 60605 312-922-7827 Hours: Monday-Thursday 9:30-4:30; Friday 9:30-9:00; Saturday and Sunday 9:30-4:30. Adults $ 4.00; children 4-17 $ 2.00; children 3 and under free; senior citizens $ 2.00. Tuesday is free.
Admission includes the sky show in the Sky Theater. The Adler Planetarium also has three floors of exhibits on astronomy and space science.

BROOKFIELD ZOO
3300 S Golf Rd, Brookfield, IL 60513 708-485-0263 Memorial Day-Labor Day 9:30-5:30 seven days a week; Labor Day-Memorial Day 10:00-4:30. General admission: Adults $ 4.00; children 3-11 and senior citizens $ 1.50; children under 3 years free. Car parking $ 4.00. Attractions additional. Tuesday and Thursday half price admission April-September, Tuesday and Thursday free admission October-March.
Home to over 23,00 animals, Brookfield Zoo attractions include Seven Seas Panorama, Habitat Africa, The Fragile Kingdom, Tropic World, Children's Zoo, Motor Safari, Aquatic Bird House, Reptile House, Pachyderm House, Australia House and Discovery Center.

CAMP SAGAWAU
Located 100 yards east of Archer Ave on Route 83 near Lemont, IL 708-257-2045 (Forest Preserve District of Cook County 708-771-1330)
This environmental education center was established to promote the study of nature and is open only for scheduled programs including workshops, college credit courses, outdoor nature photography, naturalist guided walks, field trips for adult groups, college and high school classes, and a winter Nordic ski program.

CERNAN EARTH AND SPACE CENTER
2000 N 5th Ave, River Grove, IL 60171 708-456-5815 Hours: Monday 9:00-5:00; Tuesday-Thursday 9:00-9:00; Friday 9:00-10:00; Saturday 1:00-10:00; Sunday 1:00-4:00. Adults $ 5.00; children and seniors $ 2.50.
This planetarium has sky shows, exhibits, and educational programs.

CHICAGO ACADEMY OF SCIENCES -THE NATURE MUSEUM
2001 N Clark St, Chicago, IL 60614 312-549-0606 Hotline information 312-871-2668. Hours: 10:00-5:00 seven days a week. Closed Christmas Day. Adults $ 2.00; children 17 and younger and senior citizens $ 1.00; Monday free.
The Chicago Academy of Sciences, The Nature Museum, was Chicago's first museum, founded in 1857. Visiting the Museum is like going on an indoor wilderness trail that chronicles Chicago's early geologic past.

66 Science Fun in Chicagoland

CHICAGO BOTANIC GARDEN
P. O. Box 400, 1000 Lake-Cook Road (at Edens Expressway), Glencoe, IL 60022   708-835-5440   Grounds hours: 8:00 a.m.-Sunset everyday except Christmas Day. Parking $ 4.00 per car includes admission to grounds. Tram Tickets: Adults $ 3.50, reduced rates for children, seniors and members.

The Chicago Botanic Garden has created display gardens on its 300 acres that have influenced the development of many gardens throughout the world. Home of the Chicago Horticultural Society. The Chicago Botanic Garden includes a plant information service, The Garden Shop, Food for Thought Cafe, and a library.

THE CHICAGO CHILDREN'S MUSEUM
North Pier Chicago, 435 E Illinois St, (Moving to 600 E Grand Ave, May, 1995), Chicago, IL 60611   312-527-1000   Extended summer hours: Monday-Friday 10:00-4:30; Saturday and Sunday 10:00-4:30; Friday extended 5:00-8:00; free family night Thursday 5:00-8:00. Adults $ 3.50; children and seniors $ 2.50. Ask about family memberships.

For the young child the world is a science experiment. Here the child can explore exhibits and learn at the Stinkin' Truth About Garbage, the Art and Science of Bubbles, Touchy Business, the Brainiac, the Tactile Tunnel, the Inventing Lab, and the Waterworks. Ask for a copy of the calendar of special events and workshops scheduled almost every day of the month.

CRABTREE NATURE CENTER
Palatine Road, Barrington, IL 60010 708-381-6592 (Forest Preserve District of Cook County 708-771-1330) Grounds hours: Daily 8:00-5:00. Building hours: Monday-Thursday 9:00-4:00; Saturday 9:00-12:00; Closed Friday. Admission free.

Operated by the Cook County Forest Preserve District, this center includes self-guided nature trails and a museum. School groups may phone ahead to schedule a presentation on nature.

DISCOVERY CENTER MUSEUM
711 N Main St, Rockford, IL 61103 815-963-6769 Hours: Tuesday-Saturday 11:00-5:00; Sunday 12:00-5:00; closed Monday. Adults $ 2.50; children and seniors $ 2.00; members and children under 2 years free admission.

The Museum has over 100 hands-on science exhibits inside the museum and even more in the outdoor science park. Ask for the educator's guide to Discovery Center listing traveling exhibits, special events, and school field trip information.

DUPAGE CHILDREN'S MUSEUM
Wheaton Park District Community Center, 1777 S Blanchard Rd, Wheaton, IL 60187 708-260-9960 Public hours for the school year: Tuesday 1:00-6:00; Wednesday 1:00-8:30; Thursday-Saturday 9:30-4:30; and Sunday 1:00-4:00. Adults $ 3.50; children $ 2.50; members and children under 2 admission free. Ask for calendar of adventures showing special activites almost every day of the

month. Learning Lab topics offered at the museum or in your classroom include Magnet Experimentation, What's Inside, Estimation Lab, No Numbers Math, Windows, GeoSpace, Kid's Design Engineering Lab.

EDWARD L. RYERSON CONSERVATION AREA
21950 N Riverwoods Rd, Deerfield, IL 60015 708-948-7750 Conservation area hours: 9:00-5:00 seven days a week; closed Thanksgiving, Christmas and New Years Day. Admission free.
Inquire about educational programs available including the Environmental Education Program.

ELGIN PUBLIC MUSEUM
225 Grand Blvd, Elgin, IL 60120 708-741-6655 Summer hours (April 15 - October 15): Tuesday-Sunday 12:00-4:00. Winter hours (October 15 - April 15): Saturday-Sunday 12:00-4:00 Adults $ 1.00; children 50 cents.
This natural history museum has exhibits on life science and geology. Newletter describes special events. Open to school groups Monday-Friday during the winter.

EVANSTON ECOLOGY CENTER
Evanston Environmental Association, Evanston Ecology Center, 2024 McCormick Blvd, Evanston, IL 60201 708-864-5181 Ladd Arboretum open at all times. Ecology Center hours: Tuesday-Saturday 9:00-4:30.
Ask for a copy of *Futures*, a newsletter of the Evanston Environmental Association, that describes current programs and events at the Center.

FERMILAB - GUIDED TOURS
Public Information Office, Fermilab, MS 206, P. O. Box 500, Batavia, IL 60510 Wilson Hall at Fermilab may be reached on Pine Street off Kirk Road between Butterfield Road (Rt. 56) and Wilson Street, Batavia. Accessible from Interstate 88 (East-West Tollway) Farnsworth Avenue North exit. 708-840-3351
Self-guided tours hours: 8:30-5:00 seven days a week. Pick up a self-guided tour brochure at the reception in Wilson Hall. Contact Fermilab for guided tours by appointment at 9:30 on Wednesdays and at either 9:30 or 1:30 on Thursdays and Fridays. A minimum of 10 people is needed to book a tour, with 40 being the maximum. A guided tour offers a walk through the Linear Accelerator Building. Call 708-840-3351 for more information or to make an appointment. Be sure to stop at the Lederman Science Education Center on Pine Street for the hands-on exhibit, Quarks to Quasars.

FABYAN MUSEUM
1511 S Batavia Ave, Geneva, IL 60134 708-232-4811 Open first weekend in May through second weekend in October. Hours: Tuesday, Wednesday, Thursday 11:00-3:00; Saturday and Sunday 1:00-5:00. Admission free.
This museum is the home of the late Col. George Fabyan, pioneer in acoustics and

inventor of the tuning fork. Also on this site is the Riverbank Accoustical Laboratories operated by IIT (708-232-0104) and the Riverbank Laboratory, Tuning Fork Section (708-232-2207).

THE FIELD MUSEUM
Roosevelt Road at Lake Shore Drive, Chicago, IL 60605-2497 312-922-9410 Information at extension 884. Hours: 9:00-5:00 seven days a week. Closed Thanksgiving, Christmas and New Years Day. Adults $ 5.00; children, senior citizens, and students with ID card $ 3.00; teachers and military personnel in uniform free. Wednesday free.
This great Museum of Chicago is a world class showplace of the natural science of the Earth and of the diverse cultural history of humanity. DNA to Dinosaurs, Plants of the World, Africa, Traveling the Pacific, What is We Animal, Precious Gems, and Inside Ancient Egypt are examples of exhibits. Always plan to make a return visit to see what you missed.

FIELD MUSEUM OF NATURAL HISTORY - VISITOR PROGRAMS
The Field Museum, Roosevelt Road at Lake Shore Drive, Chicago, IL 60605-2497 312-922-9410, ext 658
Contact Weekend Programs and Activities 312-922-9410, extension 288 or 350.

FULLERSBURG WOODS ENVIRONMENTAL EDUCATION CENTER
3609 Spring Rd, Oak Brook, IL 60521 (Forest Preserve District of DuPage County 708-790-4900) Hours: 9:00-5:00 seven days a week. Admission free
Programs include field trip opportunities, Learn to Be a Nature Detective, Trees Please!, From Grass to Hawks, Migration Headache, Dig Dem Bones, Maple Syrup Program, Animals in Danger. Ask for 40 page booklet, *Let's Have Class Outside Today: A Teacher's Guide to the Forest Preserve District of DuPage County.*

GARFIELD PARK CONSERVATORY
300 N Central Park Blvd, Chicago, IL 60624 312-533-1281
One of the most beautiful botanical gardens under glass in the world. Its four acres include the Palm House, the Aroid House, the Fernery, Horticultural Hall and Show House, and the Cactus House. Phone ahead for free school class tours.

THE GROVE
1421 Milwaukee Ave, Glenview, IL 60025 708-299-6096 Hours: Monday-Friday 8:00-4:30; Saturday and Sunday 9:00-5:00. Kennicott House: Sunday 1:00-4:00. Admission free.
This national landmark administered by the Glenview Park District is the 1836 homestead of the the Kennicott family. The Grove is a public museum and nature preserve. Ask for brochures about festivals and fairs held each year, animal programs that visit schools, as well as educational programs, including the Kindergarten Program, Animal Habits and Habitats, Earth Science, Pioneer Skills, Insects, Archaeology, and Pond Life.

MUSEUM OF HOLOGRAPHY /CHICAGO
1134 W Washington Blvd, Chicago, IL 60607   312-226-1007   Hours: Wednesday-Sunday 12:30-5:00. Tour groups by appointment Monday and Tuesday. Weekdays: Adults $ 3.00; school groups $ 2.00 each. Weekends and evenings: Adults $ 3.50; school groups $ 3.00 each.

This institution is dedicated to display, promote and encourage the advancement of holography as an art form. The museum displays holographic three dimensional images in various exhibition rooms. This holographic art is often the result of new research and technology produced at the School of Holography associated with the Museum.

The Children's Gallery at the Chicago Academy of Sciences can be a fun place for hands-on science activities like making bubbles. Photo courtesy of the Chicago Academy of Sciences.

70 Science Fun in Chicagoland

INTERNATIONAL MUSEUM OF SURGICAL SCIENCE
International College of Surgeons, 1524 N Lake Shore Drive, Chicago, IL 60610 312-642-6502 Tuesday-Saturday 10:00-4:00; Sunday 11:00-5:00; Closed Monday. Admission free.
This museum exhibits the methods and the history of surgery from various cultures world wide. Phone ahead for class visits to the museum. Inquire about The MED Project, a summer program for Chicago high school students.

THE JOHN G. SHEDD AQUARIUM
1200 S Lake Shore Drive, Chicago, IL 60605 312-939-2276 Hours: 9:00-6:00 seven days a week. Closed Christmas Day and New Year's Day. Adults 12-64 $ 8.00; children 3-11 $ 6.00; senior citizens over 65 $ 6.00; and children under 3 free. Admission includes Oceanarium. Half price on Thursday.
As the largest indoor aquarium in the world the John G. Shedd Aquarium has more than 6,000 aquatic animals in natural habitat exhibits. Do not miss the new Oceanarium that has its own indoor nature trails.

JOHN G. SHEDD AQUARIUM - TRIPS, EVENTS & TRAVEL
Contact Jean Majewski, Education Department, John G. Shedd Aquarium, 1200 S Lake Shore Drive, Chicago, IL 60605 312-939-2426
Local trips, special events, travel and college credit courses are available.

JURICA NATURE MUSEUM
Contact Fr. Theodore Suchy, Scholl Science Center, Illinois Benedictine College, 5700 College Rd, Lisle, IL 60532 708-960-1500, ext 1515 May 16-July 31: Wednesday 1:00-3:00; Sunday 2:00-4:00. September 1-May 15: Monday-Thursday 1:00-5:00; Friday 1:00-4:00; Sunday 2:00-4:00. Closed August. Open other times by appointment. Admission free, donation accepted.
Visitors experience the African savanna, tropical rain forest, the woodlands, wetlands and praire of Northern Illinois along with other smaller animal habitat dioramas. Insects, fish, reptiles and all kinds of birds are displayed along with many fossil specimens and skeletons. Loan program available to teachers.

KLINE CREEK FARM
County Farm Road (One-half mile north of Geneva Road), West Chicago, IL (Forest Preserve District of DuPage County 708-790-4900)
Seasonal program offerings include Fall Harvest, Maple Sugaring, Spring Planting, and The Kitchen Garden. Ask for 40 page booklet, *Let's Have Class Outside Today: A Teacher's Guide to the Forest Preserve District of DuPage County.*

KOHL CHILDREN'S MUSEUM
165 Green Bay Rd, Wilmette, IL 60091 708-256-6056 Hotline 708-251-7781 Hours: Tuesday-Saturday 10:00-5:00; Sunday 12:00-5:00; Closed Mondays. Adults and children $ 3.00; senior citizens $ 2.50; children under 1 free; Members free.

Chapter 05 - Excursions 71

For the young child the world is a science experiment. Here the child can explore and investigate hands-on exhibits including Recycle, Orientation Space, and the Technology Center that includes Bubbles Room, Fiber Optics, a walk through kaleidoscope, Computer Graphics, and more. Ask for the calendar of special events listing activities scheduled almost every day of the month. One activity is Science Club.

LAKE COUNTY FOREST PRESERVES
2000 N Milwaukee Ave, Libertyville, IL 60048-1199   708-948-7750
Ask for *Group Environmental Education Program Guide* including information about educational programs, prices, reservations, locations, dates, self-guided programs, and naturalist-guided programs. Over 20 programs available.

LEON M. LEDERMAN SCIENCE EDUCATION CENTER
- PRECOLLEGE EDUCATION PROGRAMS
Fermi National Accelerator Laboratory, P. O. Box 500   MS 777, Batavia, IL 60510   708-840-8258
Contact this center for current information about programs for high school teachers and students, programs for elementary and midlevel schools, programs for students of all ages, and science materials programs - in total over 50 programs of science education opportunities. This Science Education Center is equipped with a technology classroom, classroom laboratory, and hands-on exhibits dedicated to physical science concepts related to Fermilab. Science toys are available in a small retail shop.

LILACIA PARK
Lombard Park District, 150 S Park (Maple and Park), Lombard, IL 60148   708-627-1281   Park hours: 9:00-9:00 seven days a week. Park District hours: Monday-Friday 8:30-5:00. Park admission free, except first two weeks in May at Lilac Time. Lilac Time admission: Adults $ 1.50; senior citizens $ 1.00; children 50 cents.
This botanical garden contains eight acres of lilacs.

LINCOLN PARK CONSERVATORY
2400 N Stockton Drive at Fullerton Parkway, Chicago, IL 60614   312-294-4770
Hours: 9:00-5:00 seven days a week. Admission free.
Ask for brochure that includes information about the Palm House, the Fernery, the Cactus House, outdoor gardens, and flower shows held each year.

LINCOLN PARK ZOOLOGICAL GARDENS
2200 N Cannon Drive, Chicago, IL 60614   312-294-4660   Hours: 9:00-5:00 every day. Admission free.
The Lincoln Park Zoo is home to more than 1,600 animals, birds and reptiles from every corner of the globe. See the Great Ape House, the Bird House, the Mammal Area, the Lion House, the Penguin & Seabird House, the Primate House, the

72 Science Fun in Chicagoland

Antelope & Zebra Area, Koala Plaza, Farm-in-the-Zoo, and the Children's Zoo with baby animals. Over 25 different programs are presented for visitors each day at various times throughout the Zoo.

LITTLE RED SCHOOL HOUSE NATURE CENTER
Willow Springs Road at 104th Ave, Willow Springs, IL 60480   708-839-6897 (Forest Preserve District of Cook County 708-771-1330)   Hours: Monday-Thursday 9:00-4:30; Saturday and Sunday 9:00-5:00; closed Friday. Admission free.
Ask for brochure describing programs and events held thoughout each month. Topics include birdwatching, nature walks, astronomy, and archeology.

LIZZADRO MUSEUM OF LAPIDARY ART
220 Cottage Hill Ave, Elmhurst, IL 60126   708-833-1616   Hours: Tuesday-Saturday 10:00-5:00; Sunday 1:00-5:00; closed Monday and major holidays. Adults $ 2.50; seniors $ 1.50; children under 13 free admission.
This museum exhibits art forms made from minerals and rocks. Lapidary art is an example of how art and the science of geology combine to reveal nature's beauty.

THE MORTON ARBORETUM
Route 53 (just north of interstate 88), Lisle, IL 60532   708-719-2400   Hours: 7:00-7:00 seven days a week. Admission $ 6.00 per car; Wednesdays half-price admission $ 3.00 per car.
The Arboretum has more than 3,000 kinds of woody plants from around the world over 1,500 acres, 12 miles of roads, and 25 miles of trails. The Arboretum includes an information building, restaurant and coffee shop, gift shop, and orientation theatre.

MOTOROLA MUSEUM OF ELECTRONICS
Contact Sharon Darling, Director, Motorola, Inc., 1297 E Algonquin Rd, Schaumburg, IL 60196   708-576-6559   Hours: Monday-Friday 9:00-4:30; Closed on holidays observed by Motorola, Inc., and the first two weeks in January. The museum is open to the public by appointment by telephoning 708-576-8620. Admission free.
This museum provides interactive exhibits which chronicle the revolution in electronics technology over the 20th Century. More than 3,000 samples of Motorola products, marketing materials, and memoriabilia are housed at the museum.

THE MUSEUM OF SCIENCE & INDUSTRY
57th St and Lake Shore Drive, Chicago, IL 60637   312-684-1414   Open Memorial Day through Labor Day: 9:30-5:30 seven days a week. Winter hours: Monday-Friday 9:30-4:00; Saturday, Sunday and holidays 9:30-5:30. Closed Christmas Day. Adults $ 6.00; children 5-12 $ 2.50; senior citizens $ 5.00; and children under 5, Illinois school groups, and Park District camps

free. Omnimax Theater separately: Adults $ 6.00; children $ 2.50; senior citizens $ 5.00. Combination ticket to the Museum and Omnimax: Adults $ 10.00; children $ 5.50; senior citizens $ 8.00. Thursday free admission to the Museum.

This world class museum is famous for its numerous quality exhibits on science and technology, including the spectacular Omnimax Theater that puts the audience close to the action. It would be difficult to see everything in one day, so plan for a return visit.

MUSEUM OF SCIENCE & INDUSTRY - CURIOSITY PLACE
Contact Dori Jacobsohn, Early Childhood Specialist, Museum of Science & Industry, 57th St and Lake Shore Drive, Chicago, IL 60637 312-684-1414, ext 2428 Museum admission includes Curiosity Place. Age: Preschool

The Curiosity Place is for preschoolers age 3 to 5 years and allows them to have their own special experience with interactive exhibits designed just for them. Parents must accompany their children and appointments may be necessary at busy times.

NORTH PARK VILLAGE NATURE CENTER
5801 N Pulaski Rd, Chicago, IL 60646 312-744-5472 Hours: 10:00-4:00 seven days a week; closed holidays. Admission free.

The only nature center in the City of Chicago. The Center publishes *Urban Naturalist*, a newsletter with a calendar of events with activities almost every day of the month, pre-school registration information, school field trip descriptions and deadlines, teacher training program information, and a request for volunteers to work at the Center.

NORTHERN ILLINOIS UNIVERSITY ANTHROPOLOGY MUSEUM
Stevens Building, Northern Illinois University, DeKalb, IL 60115-2854 (Moving to new facilities in 1995 as a museum of natural history.) 815-753-0230 or 815-753-0246 Hours: Monday-Friday 9:00-5:00; closed weekends and holidays. Admission free.

This museum contains over 5,000 ethnographic objects, and specializes in cultures in Southeast Asia, New Guinea, and the Southwest and Plains Native Americans. Other collections are from Africa, modern Greece, Mesoamerica, and South America, totaling over 50,000 skeletal specimens. The Museum maintains a research library of 1,000 books and journals.

OAK PARK CONSERVATORY
615 Garfield St, Oak Park, IL 60304 708-386-4700 Hours: Monday 2:00-4:00; Tuesday-Sunday 10:00-4:00. Admission free.

This conservatory includes a cactus section, tropical plants, as well as seasonal floral shows. Inquire about current educational programs.

74 Science Fun in Chicagoland

ORIENTAL INSTITUTE MUSEUM
University of Chicago, 1155 E 58th St, Chicago, IL 60637   312-702-9521
Tuesday, Thursday, Friday, Saturday 10:00-4:00; Wednesday 10:00-8:30; Sunday 12:00-4:00. Admission free.
The Museum exhibits the cultural history of the ancient Middle East.

PILCHER PARK - JOLIET PARK DISTRICT
Contact Jerry Olson, Pilcher Park, Route 30 near I-80, Joliet, IL   815-741-7277   Hours: Monday-Friday 9:00-4:30; Saturday-Sunday 10:00-4:30
This nature center includes trails for hiking and biking, a greenhouse, and an artesian well.

THE POWER HOUSE
Contact Michael J. Radziewicz, Energy Education Administrator Commonwealth Edison, 100 Shiloh Blvd, Zion, IL 60099 708-746-7080 Hours: Monday-Saturday 10:00-5:00. Closed Sunday. Admission free.
This interactive, hands-on science museum is a fun, educational experience where one learns about energy. The exhibits focus on The Nature of Energy, The Sources and Forms of Energy, Energy Use Through Time, and Energy in Transition. It includes a theater and educational resource center.

PREHISTORIC LIFE MUSEUM
Privately owned. Within Dave's Rock Shop, 704 Main St, Evanston, IL 60202 708-866-7374   Hours: Monday, Tuesday, Thursday, and Friday 10:30-5:30; Saturday 10:00-5:00; closed Wednesday and Sunday. Admission free.
This private collection of fossils date back 1.5 billion years for algae samples and 600 millions years for fossils and has been collected over a 35 year period. Housed in one room, this collection of Earth's life history includes dinosaur eggs and nests.

RIVER TRAIL NATURE CENTER
3120 W Milwaukee Ave, Northbrook, IL 60062 708-824-8360 (Forest Preserve District of Cook County 708-771-1330) November-February hours: 8:00-4:30 seven days a week; museum open 9:00-4:00 seven days a week. March-October hours: Monday-Friday 8:00-5:00; Saturday and Sunday 8:00-5:30; museum open Monday-Thursday 9:00-4:30, Saturday and Sunday 9:00-5:00, closed Friday. Admission free.
At the Center naturalists present 15 minute presentations on nature topics. Center exhibits include native mammals, fish, amphibians, and reptiles. There are three self-guiding nature trails. November through March nature programs go out to any school in Cook County. A sugar maple festival is held each spring in March and a honey festival is held each fall in October.

SAND RIDGE NATURE CENTER
Paxton Avenue, two blocks north of 159th St, South Holland, IL 60473 708-868-0606 (Forest Preserve District of Cook County 708-771-1330)
This Center includes three nature trails, a vegetable garden with unusual vegetables and herbs, and an exhibit building.

SCITECH CAMP FOR ELEMENTARY CHILDREN
Contact Beth A. Wiegmann, Department of Curriculum and Instruction, Northern Illinois University, DeKalb, IL 60115 815-753-9025
Summer science camp for children grades 3-6 and workshops for inservice teachers.

SCITECH
- THE SCIENCE AND TECHNOLOGY INTERACTIVE CENTER
18 W Benton, Aurora, IL 60506 708-859-3434 Hours: Wednesday, Friday, Sunday 12:00-5:00; Thursday 12:00-8:00; Saturday 10:00-5:00. Adults $ 4.00; children under 18 $ 2.00; students with ID $ 2.00; family $ 8.00; Thursday evening 5:00-8:00 free. Age: preschool through adult.
This interactive science center has more than 200 hands-on exhibits on two floors demonstrating physical concepts from science and technology. At SciTech you will have an opportunity to satisfy your curiosity by enjoying scientific exploration and experimentation. SciTech's Discovery Shop has many fun science toys and materials. "Nobody flunks museum." -Frank Oppenheimer

SIX FLAGS GREAT AMERICA PHYSICS DAYS
Contact Lisa Ignoffo, Special Events Representative, Six Flags Great America, P. O. Box 1776, 542 N Route 21, Gurnee, IL 60031 708-249-2133 ext. 6439
Each spring in May, Six Flags Great America hosts Physics Days. This successful program brings high school students from all over Chicagoland to the amusement park for educational fun in a recreational atmosphere. On the rides students measure their acceleration, horsepower, and centripetal force as they become a moving part of science experiments.

SPRING VALLEY NATURE SANCTUARY
1111 E Schaumburg Rd, Schaumburg, IL 60194 708-980-2100 Hours: 9:00-5:00 seven days a week; closed Thanksgiving, Christmas and New Year's Day. Admission free.
Field trips are popular at the Sanctuary. A small museum contains dioramas about the Sanctuary as well as a small pond, a bee keeper and an aquarium with fish. Sanctuary highlights include a nature observation building, Heritage Farm, Merkle Cabin Historical Museum, Bob Link Arboretum, Illinois Heritage Grove, Spring Wildflower Display Area, Illinois Habitat Self-guided Interpretive Trail. Inquire about the Autumn Festival, the Spring Valley Environmental Education Outreach Program, and Group Venture Program.

76 Science Fun in Chicagoland

THE TIME MUSEUM
Clock Tower Inn, 7801 E State St, (Interstate 90 and business highway 20), Rockford, IL 61125 815-398-6000 Hours: Tuesday-Sunday 10:00-5:00; closed Monday. Adults $ 3.00; students $ 1.50.
This museum is an extraordinary museum of timekeeping, from Stonehenge to the Atomic Clock.

TRAILSIDE MUSEUM
738 Thatcher Ave, at Chicago Ave, River Forest, IL 60305   708-366-6530 (Forest Preserve District of Cook County 708-771-1330) Hours: 10:00-4:00 every day; except closed on Thursdays. Open 8:00-4:00 for injured animals. Admission free.
This natural history museum has no nature trails. Group reservations must be made by calling the Museum.

WALTER E. HELLER NATURE CENTER
2821 Ridge Rd, Highland Park, IL 60035   708-433-6901   Hours: Monday-Saturday 8:30-5:00; Sunday 10:00-4:00. Park hours: 6:00-9:00 seven days a week. Admission free.
This Center is a 97 acre forest with more than three miles of marked trails, including a building with a community room, a classroom, and a reference library. Request a *Park District of Highland Park Catalog* listing 80 pages of special events, educational nature classes, programs for school groups preschool through high school, and athletic programs.

WILLOWBROOK WILDLIFE CENTER
Willowbrook Forest Preserve, Park Blvd between Roosevelt and Butterfield Roads, Glen Ellyn, IL 60137   (Forest Preserve District of DuPage County 708-790-4900) Hours: 9:00-5:00 seven days a week; except Thanksgiving, Christmas Eve, Christmas and New Year Day. Admission free.
Educational programs include Young Explorers, Willowbrook Safari, Tracking, Wildlife Habitat and Survival, Sensory Awareness, Birds of Prey, Half-day Programs, and Outreach Programs. Ask for 40 page booklet, *Let's Have a Class Outside Today: A Teacher's Guide to the Forest Preserve District of DuPage County*.

WOOD LIBRARY - MUSEUM OF ANESTHESIOLOGY
Contact Dr. George Bause, Curator, 520 N Northwest Hwy, Park Ridge, IL 60068 708-825-5586 Monday-Friday 9:00-4:45. Admission free.
The American Society of Anesthesiologists' Wood Library includes a museum of equipment from the medical history of anesthesiology.

Chapter 06

Groups

Science Groups Reference Book

TRIANGLE COALITION FOR SCIENCE AND TECHNOLOGY EDUCATION - MEMBER AND AFFILIATE CONTACT DIRECTORY
Triangle Coalition for Science and Technology Education, 5112 Berwyn Rd, College Park, MD 20740-4129 301-220-0870 1994 219 pages $ 10.00
Over 100 associations, societies, academies, councils, corporations and alliances are described under the categories of Science and Engineering; Business, Industry and Labor; Education; and Affiliated Local Alliances.

Local Science Groups

ACCESS 2000
Contact Eric Hamilton, Ph.D., Director, ACCESS 2000, Loyola University of Chicago, 6525 N Sheridan, Chicago, IL 60626   312-508-3582
ACCESS 2000 is a partnership of institutions in Chicago, including universities, academies and science laboratories, that participate in programs promoting full access to mathematics, science and engineering by the year 2000 for precollege students. Contact the ACCESS 2000 office for current program opportunities in over forty different programs in the Chicago area!

AMERICAN CHEMICAL SOCIETY - CHICAGO SECTION
7173 N Austin, Niles, IL 60714   708-647-8405
This group supports the Chicagoland members of the American Chemical Society and provides information about publications and programs in chemical education. Each year this group sponsors the American Chemical Society Scholarship Examination for high school students.

BIO WEST
For information contact Fermi National Accelerator Laboratory, Lederman Science Center, P. O. Box 500   MS 777, Batavia, IL 60510   708-840-8258
Network of high school biology teachers with monthly meetings and newsletter.

CHEM WEST
For information contact Fermi National Accerator Laboratory, Lederman Science Center, P. O. Box 500   MS 777, Batavia, IL 60510   708-840-8258
Network of high school chemistry teachers with monthly meetings and newsletter.

CHICAGO ASTRONOMICAL SOCIETY
P. O. Box 30287, Chicago, IL 60630-0287
Founded in 1862, this group is America's oldest astronomical society. Anyone interested in astronomy and telescopes may enjoy this group that meets each month at the Adler Planetarium. Free telescope viewing is available each month at various locations. Ask for a copy of the society's newsletter, Cosmic Quarterly, that includes a membership application.

CHICAGO ROCKS & MINERALS SOCIETY - CHICAGO, IL
Contact Therese Donatello, President, 5252 W Palmer St, Chicago, IL 60639
One of 32 Illinois mineralogical and geological societies. Meetings at 8:00 p.m., 2nd Saturday, Sauganash Community Youth Bldg, 4600 W Peterson, Chicago, IL (September-June).

CHICAGOLAND SKY LINERS KITE CLUB
Contact Tom and Leora McCune, Chicagoland Sky Liners Kite Club, 981 Twisted Oak, Buffalo Grove, IL 60089 708-537-7066
The 11th annual Sky Circus was held at Schaumburg, Illinois, in 1994. For current information about kite festivals in the U. S. and internationally see the current KiteLines magazine for Pocket Kite Calendar and Almanac.

DES PLAINES VALLEY GEOLOGICAL SOCIETY - NILES, IL
Contact Wayne Rittorno, President, 1001 Bear Paw Ct, Carol Stream, IL 60188-9102
One of 32 Illinois mineralogical and geological societies. Meetings at 8:00 p.m., 3rd Thursday, September-June, Public Serivce Room, Our Lady of Ransom Catholic Church, 8300 N Greenwood, Niles, IL.

EARTH SCIENCE CLUB OF NORTHERN ILLINOIS - GLEN ELLYN, IL
Contact Thomas D Peterson, President, 406 Ridge Ave, Clarendon Hills, IL 60514.
One of 32 Illinois mineralogical and geological societies. Meetings at 8:00 p.m., 2nd Friday except March, July, August & October, College of DuPage, Building K, Room 157, 22nd & Lambert Rd, Glen Ellyn, IL.

EDGEWATER, UPTOWN, ROGERS PARK SCIENCE CLUBS
Contact Bryan Wunar, Loyola University Chicago, 6525 N Sheridan Road, Chicago, IL 60626 312-508-8383 Age: Elementary School through High School
Over twenty science clubs in schools and public housing provide students and young people fun with science.

ELGIN ROCK & MINERAL SOCIETY - ELGIN, IL
One of 32 Illinois mineralogical and geological societies. Meetings at 7:30 p.m., 3rd Friday, September-June, Bethlehem Lutheran Church, 340 Grand Ave, Elgin, IL.

ENVIRONMENTAL EDUCATION ASSOCIATION OF ILLINOIS
Science Education Center, 47 Horrabin Hall, Western Illinois University, Macomb, IL 61455 Student $ 6.00, Regular $ 10.00 membership fees.
This group sponsors environmental education conferences, workshops, and promotes educational curricula like Project Wild, Project Learning Tree, and CLASS Project.

GREATER OAKLAWN DIGGERS - OAKLAWN, IL
Contact Earl Machart, Jr., President, 8604 S Moody, Burbank, IL 60459-2538
One of 32 Illinois mineralogical and geological societies. Meetings at 7:30 p.m., 1st Friday, September-June, Oakview Center, Room 23, Oaklawn, IL.

80 Science Fun in Chicagoland

ILLINOIS ASSOCIATION OF BIOLOGY TEACHERS (IABT)
Contact Phil McCrea, Regional Membership Coordinator New Trier High School, 385 Winnetka Ave, Winnetka, IL 60093 708-446-7000 $ 5.00 per year.
One of the largest state affiliates of the National Association of Biology Teachers, this group has quarterly meetings and a newsletter.

ILLINOIS AUDUBON SOCIETY
34 W 269 White Thrn Rd, Wayne, IL 708-584-6290
Society dedicated to wildlife preservation.

ILLINOIS COMPUTING EDUCATORS
Affiliate of International Society for Technology in Education, c/o Vicki Logan, 137 Foster, Roselle, IL 60172 708-960-1137 Membership $ 25.00 per year.
Membership includes programs, meetings, newsletter, network for resources, and Public Domain software. Various chapters of this organization meet in the Chicagoland area. Request a membership application.

ILLINOIS JETS
Contact David L Powell, Director, Illinois JETS, University of Illinois at Urbana-Champaign, College of Engineering, 1304 W Green St, Room 207, Urbana, IL 61801-2982 800-843-5410
Annually Illinois JETS (Junior Engineering Technical Society) conducts written exam competition as well as engineering design competition in Chicago. Ask for current information and a copy of the Engineering Library for Pre-College Students and Teachers catalog of brochures, books, and videotapes. Illinois JETS has offered programs since 1959.

THE ILLINOIS SCIENCE OLYMPIAD
c/o Jeremy Way, 505 S Mathews Ave, Box 57-1, Urbana, IL 61801-6317 217-337-6582
Ask to be placed on the mailing list to receive a newsletter. The goal of this organization is to improve the quality of science education. This is accomplished through classroom activities and the encouragement of tournaments where student teamwork accomplishes scientific tasks in all areas of applied science, including biology, chemistry, physics, earth sciences and engineering. Divisions include A1 (grades K-3), A2 (grades 3-6), B (grades 5-9), and C (grades 9-12).

ILLINOIS SCIENCE TEACHERS ASSOCIATION (ISTA)
Contact George Zahrobsky, Membership Chair, Illinois Science Teachers Association, P. O. Box 2800, Glen Ellyn, IL 60138-2800 Regular membership $ 20.00 per year.
This association of Illinois science teachers K-12 publishes *Spectrum, Journal of the Illinois Science Teachers Association* (Quarterly) and in the fall the annual ISTA Convention has major speakers, group sessions, workshops, and extensive commerical exhibits.

This physics teacher is sharing teaching ideas with colleagues at a monthly meeting of the Illinois State Physics Project. All teachers and students of physics are welcome to attend these popular meetings.

ILLINOIS STATE PHYSICS PROJECT (ISPP)
Contact the Physics Department, Illinois Institute of Technology, Chicago, IL 60616-3793 312-567-3375

This friendly group of high school and college physics teachers meets each month of the school year at various school sites in the metropolitan area to share new teaching ideas and questions about the subject of physics. All teachers and students of physics are welcome to attend. A newsletter is mailed each month summarizing the last meeting, announcing the next meeting, and including news items for physics teachers.

INVENTOR'S COUNCIL
Contact Don Moyer, President, Inventor's Council, 431 S Dearborn 705, Chicago, IL 60605 312-939-3329

The Inventor's Council holds inventor's workshops on one Saturday morning each month at the Harold Washington Library Center, 400 S State St, 4th Floor NW Meeting Room, Chicago, Illinois. Topics include: How to Evaluate Patents, Patent Sucess Stories, How to Get the Best Patent, Finding the Best Way to Get New Products Made, How to Get Manufacturers to Invest in Inventions. This not-for-profit Council asks for contributions and supplies write-ups on various topics.

LAKE COUNTY GEM & MINERAL SOCIETY - WAUKEGAN, IL
Contact Duane Diehl, President, 2214 W Witchwood Ave, Lindenhurst, IL 60046.

One of 32 Illinois mineralogical and geological societies. Meetings at 7:00 p.m., 3rd Tuesday, September-June except October & December, Waukegan Public Library, Waukegan, IL.

THE LINCOLN PARK ZOOLOGICAL SOCIETY
2200 N Cannon Drive, Chicago, IL 60614 312-935-6700

Membership in the Society includes the Zoo Guidebook, Zoo Review magazine, discounts on educational programs, invitations to picnics, parties, and previews as well as free parking.

MID-LEVEL NETWORK
For information contact Fermi National Accelerator Laboratory, Lederman Science Center, P. O. Box 500 MS 777, Batavia, IL 60510 708-840-8258

Network of middle and junior high school science teachers with monthly meetings and newsletter.

MIDWEST CONSORTIUM FOR MATHEMATICS AND SCIENCE EDUCATION - NATIONAL NETWORK OF EISENHOWER MATHEMATICS AND SCIENCE
Contact Barbara Sandall, Dissemination Coordinator, North Central Regional Educational Laboratory (NCREL), 1900 Spring Rd, Suite 300, Oak Brook, IL

60521-1480 708-218-1268 69 pages
Funded by the U. S. Department of Education, this consortium works to make connections between research and practice, among schools, and between schools and communities. Whether you are a teacher or a parent, this consortium is available to serve you by providing information, technical assistance, conferences, and products.

MIDWEST HISTORICAL RESEARCH SOCIETY - CHICAGO, IL
Contact Willy Kozeluh, President, 5050 Windsor Ave, Chicago, IL 60630
One of 32 Illinois mineralogical and geological societies. Meetings at 11:30 a.m., 4th Sunday, 2600 N Narragansett, Chicago, IL.

MUSEUM OF SCIENCE & INDUSTRY - SCIENCE CLUB NETWORK
Contact Lynda Bradford, Coordinator, Museum of Science & Industry, 57th St and Lake Shore Drive, Chicago, IL 60637 312-684-1414, ext 2427 (Mark Wagner, MSI Club Coordinator, 312-684-1414, ext 2091)
Science clubs for school age children at twenty different sites in the Chicago area are coordinated through the Department of Education at the Museum of Science & Industry. Ask for current information.

PARK FOREST EARTH SCIENCE CLUB - PARK FOREST, IL
Contact Fred Aiken, President, 219 Park Dr, Glenwood, IL 60425
One of 32 Illinois mineralogical and geological societies. Meetings at 7:30 p.m., 2nd Tuesday, September-June, Freedom Hall, Park Forest, IL.

PHYSICS NORTHWEST
c/o David Thiessen, Deerfield High School, 1915 N Waukeegan Rd, Deerfield, IL 708-432-6510
This networking group of physics teachers has regular meetings and activities sharing the common interest of physics teaching. Ask to be placed on mailing list.

PHYSICS WEST
For information contact Fermi National Accerator Laboratory, Lederman Science Center, P. O. Box 500 MS 777, Batavia, IL 60510 708-840-8258
Network of high school physics teachers with monthly meetings and newsletter.

WEST SUBURBAN LAPIDARY CLUB - ELMHURST, IL
Contact Mike Reilley, President, 4100 W 100th St, Oak Lawn, IL 70453-3518.
One of 32 Illinois mineralogical and geological societies. Meetings at 8:00 p.m., 4th Friday except June-August & December, The Abbey, 407 W St Charles Rd, Elmhurst, IL.

National Science Groups

ALLIANCE FOR ENVIRONMENTAL EDUCATION
9309 Center St, Suite 101, Manassas, VA 22110   703-330-5667
This alliance of nearly 300 member organizations is committed to keeping environmental concerns at the forefront of the national agenda.

AMERICAN ASSOCIATION FOR THE ADVANCEMENT OF SCIENCE
1333 H St, NW, Washington, DC 20005-4707   202-326-6500
This distinguished organization continues to promote science education through its programs and publications. Ask for catalog of publications.

AMERICAN ASSOCIATION OF PHYSICS TEACHERS
One Physics Ellipse, College Park, MD 20740-3845   301-345-4200
This group is composed of over 10,000 physics teachers from both the high school and college level. AAPT publishes The Physics Teacher magazine. Ask for the AAPT Products Catalog.

AMERICAN CHEMICAL SOCIETY
Education Division, 1155 16th St, NW, Washington, DC 20036   202-452-2113
Ask for 24-page *Teaching Resources Catalog* that includes posters, videotapes, magazines, educational materials, curriculum supplements, National Chemistry Examinations, and the annual Chemistry Calendar.

AMERICAN GEOLOGICAL INSTITUTE
4220 King St, Alexandria, VA 22302 703-379-2480 AGI Publications Center, P. O. Box 205, Annapolis Junction, MD 20701   301-953-1744
This federation of professional associations serves geoscientists. Ask for 6 page brochure listing publications available, including Earth Science Guidelines Grades K-12, Earth-Science Education Resource Directory, and Careers in the Geosciences.

AMERICAN RADIO RELAY LEAGUE - DEPARTMENT OF EDUCATIONAL ACTIVITIES
American Radio Relay League, 225 Main St, Newington, CT 06111   203-666-1541
This is the best resource of information about Amateur Radio for all ages. Inquire about educational materials available including video courses, Morse Code and Ham Radio materials, *Amateur Radio in the Classroom* newsletter, license publications for students, and reference books. Ask for list of School Amateur Radio Clubs and Advisors.

Chapter 06 - Groups 85

ASSOCIATION FOR WOMEN IN SCIENCE
1522 K St, NW, Suite 820, Washington, DC 20005   202-408-0742
This nonprofit educational association is dedicated to increasing the educational and employment opportunities for young girls and women in all fields of science. Ask for brochure describing publications and resources available.

FUTURE SCIENTISTS & ENGINEERS OF AMERICA (FSEA)
Contact George Westrom, Executive Director, P. O. Box 9577, Anaheim, CA 92812   714-774-5000, ext. 6010
FSEA chapters are for students in grades 4-12 and chapter meetings after school hours allow students to work on challenging science and engineering projects. Each FSEA project is conducted by a teacher and mentor team. Chapters are sponsored by business, professional societies, and community organizations.

HISTORY OF SCIENCE SOCIETY, INC.
Contact Dr. Keith Benson, History of Science Society, Inc., Department of Medical History and Ethics, University of Washington, DR-05, Seattle, WA 98195   206-543-9366
Dedicated to the study of the history of science, this society publishes the journal, *ISIS*, *The HSS Newsletter*, and *Guide to the History of Science*. Founded in 1924, the Society advances research and teaching in the history of science.

INSTITUTE OF ENVIRONMENTAL SCIENCES
940 E Northwest Hwy, Mt. Prospect, IL 60056   708-255-1561
The purpose of this professional society of engineers, scientists, and educators is to understand the effects of environmental factors as they relate to the production and safe operation of manufactured products. Areas of interest include knowledge pertaining to environmental sciences, design, contamination control, encourage courses in the environmental sciences, and to recognize related achievement. Publishes the *Journal of the IES*.

INTERNATIONAL SOCIETY FOR TECHNOLOGY IN EDUCATION
1787 Agate St, Eugene, OR 97403-1923   800-336-5191   503-346-4414
This professional organization is dedicated to the improvement of all levels of education through the use of computer-based technology. Ask for a copy of ISTE's 40 page publications catalog offering special journals, books on computers in education, and educational software for educators and the classroom.

INTERNATIONAL TECHNOLOGY EDUCATION ASSOCIATION
1914 Association Dr, Reston, VA 22091-1502   703-860-2100
This association is dedicated to promoting technological literacy. Ask for 20 page catalog of publications and classroom materials.

86 Science Fun in Chicagoland

JUNIOR ENGINEERING TECHNOLOGY SOCIETY, INC. (JETS)
1420 King St, Suite 405, Alexandria, VA 22314-2715   703-548-JETS
This nationwide organization is for precollege students (grades 9-12) interested in engineering, technology, mathematics and science. JETS activities include testing for the National Engineering Aptitude Search and the National Engineering Design Callenge. Ask for listing of career-related and other publications available from JETS-Guidance at the above address. JETS has offered programs since 1949.

NATIONAL ACTION COUNCIL FOR MINORITIES IN ENGINEERING, INC.
3 W 35th St, 3rd Floor, New York, NY 10001   212-279-2626
This nonprofit organization is dedicated to bringing the talents of African Americans, Hispanics, and American Indians to the nation's workforce.

NATIONAL ASSOCIATION FOR SCIENCE, TECHNOLOGY AND SOCIETY (NASTS)
133 Willard Building, University Park, PA 16802   814-865-9951
The mission of this association is to create a technologically and scientifically literate citizenry, to integrate the humanities and social sciences with science and technology, to help shape public policy which guides evolving technology, and to introduce science and technology to the public.

NATIONAL ASSOCIATION OF BIOLOGY TEACHERS
11250 Roger Bacon Dr, #19, Reston, VA 22090-5202   703-471-1134 $ 48.00 dues per year.
This is the only national association specifically organized to assist teachers in the improvement of biology education.   Ask for publications and membership brochures.

NATIONAL ASSOCIATION OF GEOLOGY TEACHERS, INC.
P. O. Box 5443, Bellingham, WA 98227-5443   206-650-3587
This association seeks to foster improvement in teaching earth sciences at all levels and to disseminate knowledge in this field to the general public. Ask for information about materials available.

NATIONAL ENERGY FOUNDATION
5160 Wiley Post Way, Suite 200, Salt Lake City, UT 84116   801-539-1406
This nonprofit organization provides programs and materials to help promote an awareness of energy-related issues. Ask for 15 page catalog of publications and science kits. Materials include Out of the Rock, a mineral resource and mining education program for K-8 produced in conjunction with the U. S. Bureau of Mines.

NATIONAL SCIENCE RESOURCES CENTER (NSRC)
Smithsonian Institution - National Academy of Sciences, Smithsonian Institution, Arts & Industries Building, Room 1201, Washington, DC 20560 202-357-2555
The NSRC works to improve the teaching of science in the nation's schools. NSRC disseminates information about effective science teaching resources, develops curriculum materials, and sponsors outreach and leadership development activities. Ask to be placed on the mailing list for the *NSRC Newsletter*.

NATIONAL SCIENCE TEACHERS ASSOCIATION (NSTA)
1840 Wilson Blvd, Arlington, VA 22201-3000   703-243-7100   800-722-NSTA
This association plays a major role in improving science teaching at all levels, preschool through college. Its membership is composed of 50,000 science teachers, science supervisors, administrators, scientists, and others involved directly and indirectly with science education. Ask for publications and membership brochure.

SCHOOL SCIENCE AND MATHEMATICS ASSOCIATION
126 Life Science Building, Bowling Green State University, Bowling Green, OH 43403   419-372-7393
Founded in 1901, the purpose of this association is to disseminate research findings and its implications for school practice.

THE SCIENCE OLYMPIAD
5955 Little Pine Lane, Rochester, MI 48306   313-651-4013
The goal of this international organization is to improve the quality of science education and is accomplished through classroom activities and the encouragement of tournaments at the district, regional, state, and national levels.

SCIENCE SERVICE, INC.
1719 N St, NW, Washington, DC 20036   202-785-2255
The Science Youth Program of Science Service, Inc. includes the following activities: Westinghouse Science Talent Search, and the International Science and Engineering Fair. Science Service, Inc. publishes *Science News, The Weekly Newsmagazine of Science*, and *The Directory of Student Science Training Programs*.

U. S. DEPARTMENT OF ENERGY
1000 Independence Ave, SW, Room 3F-061, Washington, DC 20585 202-586-8949
The Department of Energy has six national laboratories designated as Laboratory Science Education Centers, including Argonne, IL. These centers provide a range of college and precollege education programs. Contact Education Programs, Argonne National Laboratory, 9700 S Cass Ave, Argonne, IL 60439 at 312-972-3373 for current information about programs.

Chapter 07 Instruments

Scientific Measurement Instruments

Tools are a part of any field of knowledge. Good science comes from knowing the tools and how they are used. The following list of scientific instruments is a good beginning for the person who wants to do quality science. See Chapter 09 - Materials for sources where you can purchase these instruments.

ALTIMETER
Measures atmospheric pressure and height in units of feet.
The aneroid barometer is adapted for use in aircraft as an altimeter allowing measurement of height above the ground.

ANEMOMETER
Measures wind speed in units of miles per hour.
This instrument is also called a cup anemometer.

ANEROID BAROMETER
Measures atmospheric pressure in units of inches or millimeters of mercury.
Aneroid barometers use a partially evacuated metal chamber connected to a spring and a dial that directly displays atmospheric pressure on the dial without need for liquid mercury as in the mercury barometer.

BALANCE
Measures mass in units of grams (g).
A balance measures how much Earth's gravity pulls down on an object compared to standard mass sizes. This comparison is a measure of the object's inertia, or resistance to motion. By definition, one kilogram of water has a volume of one liter and one gram of water has a volume of one milliliter.

BAROGRAPH
Measures atmospheric pressure over time.
The barograph adapts the aneroid barometer for continuously recorded pressure changes on a moving sheet of recording paper.

ECHO SOUNDER
Measures ocean depth in units of feet and meters.
A sounding instrument, used in sea water, for automatically determining the depth of the sea floor or of an object beneath a ship.

GRADUATED CYLINDER
Measures volume in units of milliliters (ml).
Read liquid volume in a graduated cylinder in milliliters at the bottom of the liquid's curved surface, called the meniscus.

HYDROMETER
Measures density in units of grams per milliliter (g/ml).
The hydrometer floats in a liquid to measure its density. The higher the hydrometer floats, the more dense the liquid. The hydrometer is often confused with the hygrometer.

HYGROMETER
Measures relative humidity in units of percent.
The hygrometer operates on the principle that a synthetic fiber will curl with increasing air moisture or humidity. Humidity is read from a circular dial on the hygrometer. The hygrometer is often confused with the hydrometer.

MERCURY BAROMETER
Measures atmospheric pressure in units of inches or millimeters of mercury.
The mercury barometer is a long glass tube of liquid mercury inverted into an open dish of liquid mercury. The atmospheric pushes on the mercury surface in the dish forcing the mercury in the tube to a height corresponding to atmospheric pressure.

METERSTICK
Measures length in units of meters (m), centimeters (cm), millimeters (mm).
The meterstick measures one meter long which is equivalent to 100 cm, 1000 mm, and 39.37 inches. The meter was originally defined so that 10 million meters would exactly measure the distance from the equator to the north pole of the Earth.

MICROMETER
Measures length in units of .01 of a millimeter.
Dimensions of small objects can be measured with this instrument. A thimble is turned on a graduated barrel to close a spindle and a stationary anvil onto an object. The micrometer can measure the diameter of a copper wire or a human hair.

MOHS HARDNESS SCALE
Measures mineral & rock hardness on a scale from 1 - 10.
The Mohs hardness scale measures relative hardness on a scale from 1 (talc) to 10 (diamond). For example, if a mineral of hardness 4 will scratch a sample mineral, but a hardness 3 mineral will not, then your sample has a hardness of 3-4.

PH SCALE
Measures acidity (or alkalinity) on a scale from 0 - 14.
The pH scale is a logarithmic scale. A value of 7 denotes a neutral solution, values below 7 indicate greater acidity, and values above 7 indicate greater alkalinity.

PIPETTE
Measures liquid volume in units of milliliters (ml).
A calibraded glass tube for measuring small liquid volumes where the liquid is drawn into the tube by suction.

PROTRACTOR
Measures angular distance in units of degrees.
An instrument used for measuring angles on paper or in the laboratory.

RICHTER SCALE
Measures earthquake intensity on a scale from 0 - 10.
A scale of earthquake intensity based on the motion of a seismometer. An earthquake of magnitude 10 would shake the entire Earth.

SCALE OR BATH SCALE
Measures weight in units of pounds or kilograms.
A scale measures weight, the pull or force of gravity, with a spring mechanism. On the Earth's surface weight in pounds has a direct correspondence to mass in kilograms. One kilogram equals 2.2 pounds. On other planets the correspondence is different. In the metric system force or weight is measured in newtons; in the English system force or weight is measured in pounds.

SEISMOMETER
Measures earthquake vibrations.
Seismographs measure vibrational waves moving through the Earth from earthquake sources. In a seismograph a heavy weight freely suspended from a support does not move with Earth movement. The seismometer records vibrational intensity of the seimometer's surroundings, which is the Earth's vibrational movement.

SLING PSYCHROMETER
Measures relative humidity in units of percent.
The sling psychrometer is made of two thermometers with one wet and one dry bulb. The drop in temperature on a wet bulb thermometer allows the measurement of relative humidity by using a humidity data table. A larger drop in temperature would correspond to a lower humidity.

THERMOGRAPH
Measures temperature over time.
A thermograph is made from a thermometer which continuously records temperature on a moving sheet of recording paper.

THERMOMETER
Measures temperature in units of degrees centigrade (Celsius) or Fahrenheit.
The thermometer measures temperature by observing a liquid which expands with increasing molecular motion due to heat addition. The thermometer measures temperature as the amount of average molecular motion. Heat is thermal energy. The Kelvin scale is defined by gas laws and absolute zero.

VERNIER CALIPER
Measures length in units of .1 of a millimeter.
The vernier caliper is a caliper with a vernier scale which allows the measurement of length to the nearest .1 of a millimeter. The allignment of marks on the vernier scale accurately measures one more figure of length. This instrument can measure the diameter of a child's marble or a pencil.

WIND VANE
Measures wind direction in units of angular degrees.
An instrument used to determine wind direction. A design like an arrow, mounted on a vertical axis, is forced to point in the direction of the wind's motion. A north wind blows from the north to the south.

Chapter 08

Libraries

Library Directories

DIRECTORY OF SPECIAL LIBRARIES AND INFORMATION CENTERS
edited by Brigitte T. Darnay Gale Research Company, Detroit, MI 48226 Vols 1-2, Bound in 3 parts. Available at major libraries.
This extensive reference lists North American libraries, describes their collections, and indexes them by geographic regions and by science categories. This reference has it all for the information addict.

SCIENCE & TECHNOLOGY LIBRARIES
The Haworth Press, Inc., 10 Alice St, Binghamton, NY 13904-1580 607-722-5857 Quarterly $ 40.00 per year.
This periodical publishes articles written for librarians of science and technology libraries. A list of books for sale at the end of each issue, however, includes various reference books listing science and technology libraries.

Local Libraries with Science Collections

This list contains libraries which have good to fine science book collections. Many libraries specialize in various topics. Some are general collection libraries with good science collections.

Some of the libraries listed below have limited circulation privileges or are not open to the public. (Libraries not open to the public are listed below in an attempt to make this list as complete as possible.) Please contact each library for current library policies and computer modem access numbers to their catalogs if available.

Local public libraries are not included in the following list with the exception of Chicago's three major public libraries. Your local public library will have many interesting children's science books and books about science fair project ideas.

A T & T BELL LABORATORIES - INDIAN HILL LIBRARY
Contact Ruby K. Chu, Librarian, AT&T Bell Laboratories, 200 N Naperville Rd, Naperville, IL 60566 708-979-2551
This library is open to the public, but building security restrictions require telephoning for an appointment. The collection contains 30,000 books and 325 periodicals on computers, telecommunications, business and electronics.

A T & T INFORMATION RESOURCE CENTER
Contact Ruby K. Chu, Librarian, 2600 Warrenville Rd, Lisle, IL 60532 708-224-4000
Not open to the public.

ABBOTT LABORATORIES - ABBOTT INFORMATION SERVICES
Contact John D Opem, Manager, Abbott Laboratories, One Abbott Park Rd, Abbott Park, IL 60064-3500 708-937-6012
Not open to the public. This library makes books available, however, through interlibrary loan. The collection contains 25,000 books and 1300 periodicals.

ADLER PLANETARIUM - LIBRARY
Contact Dr. Evelyn D. Natividad, Library Administrator, Adler Planetarium, 1300 S Lake Shore Drive, Chicago, IL 60605 312-322-0593
This reference library is open to the public by appointment, but does not extend circulation privileges. The collection contains over 5,000 books related to astronomy.

94 Science Fun in Chicagoland

AMERICAN MEDICAL ASSOCIATION
- DIVISION OF LIBRARY AND INFORMATION MANAGEMENT
Contact Sandra Schefris, Director, American Medical Association, 515 N State St, Chicago, IL 60610 312-464-4855
Not open to the public.

AMERICAN NUCLEAR SOCIETY - LIBRARY
Contact Lois S. Webster, Executive Assistant, American Nuclear Society, 555 N Kensington Ave, La Grange Park, IL 60525 708-352-6611
This library is open to the public. The collection contains approximately 3000 books and 150 periodicals on nuclear energy, science and business management.

AMOCO CORPORATION - CENTRAL RESEARCH LIBRARY
Amoco Research Center, Box 3083, Naperville, IL 60566-7083 708-420-5543
This library is not open to the public, but books are available through interlibrary loan. The collection contains approximately 45,000 books and 2500 periodicals on science and technology.

AMOCO CORPORATION - LIBRARY/INFORMATION CENTER
Contact Vicky A. Perlman, Manager, Amoco Corporation, 200 E Randolph St, Chicago, IL 60601-7125 312-856-5961
This library is open by appointment to individuals for specific research and contains 70,000 books and documents as well as 1000 periodicals related to petrochemicals, business and law.

ARGONNE NATIONAL LABORATORY
- TECHNICAL INFORMATION SERVICES DEPARTMENT
Argonne National Laboratory, 9700 S Cass Ave, Argonne, IL 60439 708-252-4215
Argonne National Laboratory has ten different libraries at various sites within the Laboratory which are only open to individuals with clearance and admission to the Laboratory.

BROOKFIELD ZOO - LIBRARY
3300 S Golf Rd, Brookfield, IL 60513 708-485-0263
This library can be contacted for reference information.

CENTER FOR RESEARCH LIBRARIES
6050 S Kenwood Ave, Chicago, IL 60637-2804 312-955-4545 By appointment. Donald Simpson, President.
This information center is an extensive storage center for hard to find research dissertations, especially foreign dissertations. The reading room is open to the public at no cost when given 48 hours notice of specific reading need. If you know what you want and it is hard to find, this might be the place to go.

CHICAGO ACADEMY OF SCIENCES
- MATTHEW LAUGHLIN MEMORIAL LIBRARY
2001 N Clark St., Chicago, IL 60614 312-549-0606 Founded 1857.
Library temorarily closed and not open to the public.

CHICAGO BOTANIC GARDEN - LIBRARY
Contact Virginia A. Henrichs, Librarian, Chicago Botanic Garden, P. O. Box 400, 1000 Lake-Cook Road (at Edens Expressway), Glencoe, IL 60022-0400 708-835-8200 or 8201
This library is open to the public and requires a $ 4.00 parking fee for nonmembers. The collection contains 13,000 books and 300 periodicals on horticulture, botany and natural history.

CHICAGO PUBLIC LIBRARY
- CARTER G. WOODSON REGIONAL LIBRARY
Contact Emily Guss, Director, Carter G. Woodson Regional Library, 9525 S Halsted St, Chicago, IL 60628 312-747-6900
One of two major regional libraries in Chicago.

CHICAGO PUBLIC LIBRARY
- CONRAD SULZER REGIONAL LIBRARY
Contact Leah J. Steele, Director, Conrad Sulzer Regional Library, 4455 N Lincoln Ave, Chicago, IL 60625 312-744-7616
One of two major regional libraries in Chicago.

CHICAGO PUBLIC LIBRARY
- THOMAS HUGHES CHILDREN'S LIBRARY
Contact Laura B. Culberg, Head, Harold Washington Library Center, 400 S State St, Chicago, IL 60605 312-747-4200
The Hughes Children's Library contains 100,000 books for children. A majority of these books are non fiction. Look under Q within the Library of Congress system for children's science books.

CHICAGO PUBLIC LIBRARY HAROLD WASHINGTON CENTER
- BUSINESS/SCIENCE/TECHNOLOGY DIVISION
Contact David R Rouse, Divison Chief, Harold Washington Library Center, 4th Floor, 400 S State St, Chicago, IL 60605 Science Reference 312-747-4450 and Voice Mail 312-747-4470
The Central Library of the Chicago Public Library contains approximately 50,000 different book titles on pure science subjects.

96 Science Fun in Chicagoland

CHICAGO STATE UNIVERSITY
- THE PAUL AND EMILY DOUGLAS LIBRARY
The Paul and Emily Douglas Library, Chicago State University, E 95th St & King Dr, Chicago, IL 60628   Reference desk 312-995-2235, Circulation 312-995-2341
This university library is open to the public, but library privileges are restricted. The Douglas Library contains 360 science periodicals.

CHICAGO ZOOLOGICAL SOCIETY - LIBRARY
Contact Mary S. Rabb, Librarian, Chicago Zoological Society, 8400 W 31st St, Brookfield, IL 60513 708-485-0263, ext 580
This library is not open to the public, but extends interlibrary loan privileges. The collection contains 10,000 books related to zoology.

COMMONWEALTH EDISON COMPANY - LIBRARY
Contact Grace M. Pertell, Librarian, Commonwealth Edison Company, First National Plaza, 35th Floor, P. O. Box 767-35 FNE, Chicago, IL 60690 312-394-3066
This collection contains information on energy, the environment, electrical engineering, and management including 300 different journal titles. It is open to the public and requires admission through security.

DE PAUL UNIVERSITY - LINCOLN PARK CAMPUS LIBRARY
DePaul University, 2350 N Kenmore Ave, Chicago, IL 60614   312-362-6922
This university library is open to the public, but circulation privileges are restricted. The general collection of this library contains 620,000 books and 4,200 periodicals.

DE VRY INSTITUTE OF TECHNOLOGY
- LEARNING RESOURCE CENTER
Contact Julie A. Engel, Director, De Vry Institute of Technology, 3300 N Campbell Ave, Chicago, IL 60618   312-929-8500
This library is open to the public, but circulation privileges are restricted. This library contains 10,000 books and 50 periodicals.

FIELD MUSEUM OF NATURAL HISTORY - LIBRARY
Contact W. Peyton Fawcett, Head Librarian, The Field Museum, Roosevelt Road & Lake Shore Dr, Chicago, IL 60605   312-922-9410, ext 282
The reference library of the Field Museum of Natural History is open to the public, but the library does not circulate its collection. Enter via the west side of the Museum's north entrance. This collection contains over 250,000 books.

GAS RESEARCH INSTITUTE - LIBRARY SERVICES
Gas Research Institute, 8600 W Bryn Mawr Ave, Chicago, IL 60631 312-399-8386
Not open to the public.

Chapter 08 - Libraries 97

GOVERNORS STATE UNIVERSITY - LIBRARY
Governors State University, University Park, IL 60466   708-534-5000
This university library is open to the public, but circulation privileges are restricted. This library's general collection contains 300,000 books.

HONEYWELL, INC. - COMMERCIAL BLDG. GROUP - TECHNICAL LIBRARY
Honeywell, Inc., 1500 W Dundee, Arlington Heights, IL 60004   708-797-4000
This extensive collection consists of technical literature related to products. Although it is not open to the public, special permission use may be obtained.

ILLINOIS (STATE) DEPARTMENT OF ENERGY AND NATURAL RESOURCES - CHICAGO ENERGY OPERATIONS LIBRARY
Contact Alice I. Lane, Library Associate, Illinois Department of Energy and Natural Resources, State of Illinois Center, 100 W Randolph, Suite 11-600, Chicago, IL 60601   312-814-3895
This small library is not open to the public. It is available, however, for reference use only and books do not circulate.

Study pays off for these students taking a Junior Engineering Technical Society (JETS) scholarship exam.

98 Science Fun in Chicagoland

**ILLINOIS INSTITUTE OF TECHNOLOGY
- PAUL V. GALVIN LIBRARY**
35 W 33rd St, Chicago, IL 60616-3793   312-567-6844
This extensive collection of over 500,000 books on science and technology is open to the public, however, circulation privileges are restricted.

**ILLINOIS INSTITUTE OF TECHNOLOGY RESEARCH INSTITUTE
- MANUFACTURING TECHNOLOGY INFORMATION ANALYSIS CENTER**
Contact Michal Safar, Director, IIT Research Institute, 10 W 35th St, Chicago, IL 60616   800-421-0586
This center is open to the public for literature searches and contains approximately 300 books and 300 periodicals on manufacturing technology.

**INSTITUTE OF GAS TECHNOLOGY
- TECHNICAL INFORMATION CENTER**
Contact Carol Worster, Supervisor, Institute of Gas Technology, 1700 S Mt Prospect Rd, Des Plaines, IL 60018   708-768-0664
This library is open to the public and contains 33,000 books, 100,000 technical reports, and 500 periodicals on gas, energy and biomass.

JOHN G. SHEDD AQUARIUM - LIBRARY
Contact Janet E. Powers, Coordinator of Library Services, John G. Shedd Aquarium, 1200 S Lake Shore Drive, Chicago, IL 60605   312-939-2426
This library is open only to Aquarium membership and individuals granted admission for research. Contact the Aquarium about membership.

**LEON M. LEDERMAN SCIENCE EDUCATION CENTER
- TEACHER RESOURCE CENTER**
Contact Susan Dahl, Coordinator, Teacher Resource Center, Fermi National Accelerator Laboratory, Leon M. Lederman Science Education Center, P. O. Box 500 MS 777, Batavia, IL 60510   708-840-8258 Monday-Friday 8:30-5:00; Saturday 9:00-3:30. Call for an appointment.
This extensive teacher resource center is filled with books, periodicals, kits, videotapes, etc. - for teachers, administrators, librarians, scientists, and Science Center program participants. Call for an appointment.

LINCOLN PARK ZOOLOGICAL GARDENS - LIBRARY
Lincoln Park Zoological Gardens, 2200 N Cannon Dr, Chicago, IL 60614-3895
312-294-4640
This small library is open to the public, but has no circulation privileges.

Chapter 08 - Libraries   99

LOYOLA UNIVERSITY CHICAGO - SCIENCE LIBRARY
Contact Stephen Macksey, Head of the Science Library, Loyola University Chicago, 6525 N Sheridan Rd, Chicago, IL 60626   312-508-8400
This university library is open to the public, but public visitation is limited to two hours per week. Circulation privileges are restricted.

METROPOLITAN WATER RECLAMATION DISTRICT OF GREATER CHICAGO - TECHNICAL LIBRARY
Contact Andrew King, Librarian, Metropolitan Water Reclamation District of Greater Chicago, 100 E Erie St, Chicago, IL 60611   312-751-6658
This small library is open to the public for research purposes only, and has no circulation privileges. The collection is limited to works supporting scientists and engineers at the District.

MORTON ARBORETUM - STERLING MORTON LIBRARY
Contact Michael T. Stieber, Reference Librarian, Morton Arboretum, Lisle, IL 60532   708-719-2427
This library is open to the public and circulation privileges are extended to members and through interlibrary loan. The collection contains 25,000 books and 400 periodicals.

MOTOROLA, INC. - CIG/GSS LIBRARY
Contact Jennifer Mielke, Technical Librarian, Motorola, Inc., 1501 W Shure Dr, Arlington Heights, IL 60004   708-632-4133
This technical library basically contains technical research journals and is not open to the public.

MOTOROLA, INC. - COMMUNICATION SECTOR LIBRARY
Contact Marion Mason, Library Manager, Motorola, Inc., 1301 E Algonquin Rd, Room 1914, Schaumburg, IL 60196   708-576-5940
This industry library is not open to the public. An information pass will be extended, however, to researchers needing information not available elsewhere. This collection contains 8,000 books and 500 periodicals on electrical engineering.

MUSEUM OF SCIENCE & INDUSTRY - LIBRARY
Contact Pam Nelson, Librarian, Museum of Science & Industry, 57th St & Lake Shore Drive, Chicago, IL 60637 312-684-1414, ext 2449
Library open to the public by appointment. Note, this collection duplicates many public library collections. Check with your public library first.

MUSEUM OF SCIENCE & INDUSTRY - NASA TEACHER RESOURCE CENTER
Contact Ed McDonald, Museum of Science & Industry, 57th St and Lake Shore Drive, Chicago, IL 60637   312-684-1414, ext 2423
This teacher resource center is open by appointment. It contains numerous books, slides, video tapes, posters and educational materials about the United States space program.

NATIONAL SAFETY COUNCIL - LIBRARY
Contact Robert J. Marecek, Manager, National Safety Council, 1121 Spring Lake Dr, Itasca, IL 60143-3201   708-285-1121
This library is open to the public and contains a comprehensive collection of 140,000 documents on health and safety. Fees apply to researchers requiring extensive assistance. It is best to telephone ahead for an appointment.

NATURE CONNECTIONS PROJECT - CHICAGO PUBLIC LIBRARY'S HAROLD WASHINGTON LIBRARY CENTER
Contact Jane Sorensen, Thomas Hughes Room, Harold Washington Library Center, 400 S State St, Chicago, IL 60605   312-747-4633   Age: Elementary School
This project includes a collection of things from nature in book form that is available through the Chicago Public Library.

NORTHEASTERN ILLINOIS UNIVERSITY - RONALD WILLIAMS LIBRARY
Contact Bradley F. Baker, Librarian, Northeastern Illinois University, 5500 N St Louis Ave, Chicago, IL 60625-4699   312-794-2615
This university library is open to the public, but circulation privileges are restricted. Books are available, however, through interlibrary loan. The total general collection is broad and contains 380,000 books and 3,500 periodicals.

NORTHWESTERN UNIVERSITY - SEELEY G. MUDD LIBRARY FOR SCIENCE AND ENGINEERING
Contact Robert Michaelson, Head Librarian, Northwestern University, 2233 Sheridan Rd, Evanston, IL 60208-2300   708-491-3362
This library is open to the public, but circulation privileges are restricted. For guest circulation information telephone 708-491-7617. The collection contains 230,000 books and 1800 periodicals on science and engineering.

THE POWER HOUSE - ENERGY RESOURCE CENTER
Contact Mary L. Crompton, Energy Education Assistant, Commonwealth Edison, 100 Shiloh Blvd, Zion, IL 60099   708-746-7850
This science education resource center is open to both students and teachers for research and study. It contains over 500 books, 48 periodicals, six computers and four VCR's with video monitors.

QUAKER OATS COMPANY - JOHN STUART RESEARCH LABORATORIES - RESEARCH LIBRARY
Contact Geraldine R. Horton, Manager, Quaker Oats Company, 617 W Main St, Barrington, IL 60010 708-381-1980, ext 2050
This collection contains 7,000 books on food and food processing and is open to the public by appointment.

ROOSEVELT UNIVERSITY - MURRAY-GREEN LIBRARY
Contact Promilla Bansal, Head of Reference, Roosevelt University, 430 S Michigan Ave, Chicago, IL 60605 312-341-3643
This library is open to the public, but circulation privileges are restricted. The collection contains 300,000 books of which 12,000 are on science topics.

SEARLE LIBRARY
4901 Searle Pkwy, Skokie, IL 60077 708-982-8285
Not open to the public.

U. S. ENVIRONMENTAL PROTECTION AGENCY - REGION 5 LIBRARY
Contact Ms. Lou W. Tilley, Regional Librarian, U. S. Environmental Protection Agency, 77 W Jackson Blvd, 12th Floor, Chicago, IL 60604 312-353-2022
This library is open to the public and its collection is primarily EPA reports as well as 300 periodicals related to the environment.

UNIVERSITIES RESEARCH ASSOCIATION - FERMI NATIONAL ACCELERATOR LABORATORY - LIBRARY
Contact Paula Garrett, Library Administrator, Fermi National Accelerator Laboratory, Box 500, MS 109, Batavia, IL 60510 708-840-3401
This library is open to the public, but does not extend library privileges. The collection contains over 10,000 books primarily on high energy physics.

UNIVERSITY OF CHICAGO - JOHN CRERAR LIBRARY
Contact Patricia K. Swanson, Asst. Dir. for Sci. Libraries, 5730 S Ellis, Chicago, IL 60637 312-702-7715
This distinguished collection of science, technology and medicine contains approximately one million (1,000,000) books. Although this library is open to the public, library circulation privileges are not available to the public.

102 Science Fun in Chicagoland

UNIVERSITY OF CHICAGO - SPECIAL COLLECTIONS
Contact Alice Schreyer, Curator, Joseph Regenstein Library, 1100 E 57th St, Chicago, IL 60637 312-702-8705
This collection of 250,000 books and 19,000 linear feet of manuscripts includes works from science history. Reading room privideges open to the public include special rules necessary for the proper care of books hundreds of years old. Here you can see and hold the history of science in your hands.

UNIVERSITY OF ILLINOIS AT CHICAGO - SCIENCE LIBRARY
Contact Julie M. Hurd, Science Librarian, University of Illinois at Chicago, Science & Engineering South Bldg, 845 W Taylor, Chicago, IL 60607 312-996-5396
This library is open to the public, but circulation privileges are restricted. The collection contains approximately 60,000 books and 1000 periodicals on biology, chemistry, geology and physics.

ZENITH ELECTRONICS CORPORATION - TECHNICAL LIBRARY
Contact Eleanore L. Berns, Information Manager, Zenith Electronics Corporation, 1000 N Milwaukee Ave, Glenview, IL 60025 708-391-8452
This library is not open to the public.

Chapter 09

Materials

Hands-On Materials Reference Books

There are numerous sources of free and inexpensive hands-on materials for doing science activities:

EDUCATORS GUIDE TO FREE SCIENCE MATERIALS - 35TH EDITION
edited by Mary H. Saterstrom Educators Progress Service, Inc., 214 Center St, Randolph, WI 53956-1497 414-326-3126 1994 296 pages $ 27.95
This book lists and describes free science materials, including films, filmstrips, slides, audiotapes, and printed materials by category of science subject area.

ENERGY EDUCATION RESOURCES - KINDERGARTEN THROUGH 12TH GRADE
National Energy Information Center, EI-231, Energy Information Administration, Room 1F-048, Forrestal Building, 1000 Independence Ave, SW, Washington, DC 20585   202-586-8800   1992
Ask for a copy of this 31 page booklet listing 86 different sources of educational materials from both public and private institutions and companies. Each source usually offers a catalog listing free materials.

NSTA SCIENCE EDUCATION SUPPLIERS
A Supplement to *Science & Children, Science Scope*, and *The Science Teacher*, National Science Teachers Association, 1840 Wilson Blvd, Arlington, VA 22201-3000   800-722-NSTA Published annually. 122 pages $ 5.00 per copy.
List of science educational materials manufacturers and distributors. The most current and comprehensive list of manufacturers, publishers and distributors of science education materials. See Equipment/Supplies. In 1994, 275 sources of hands-on equipment and supplies were listed.

THE PLASTICS AND THE ENVIRONMENT SOURCEBOOK
The Polystyrene Packaging Council, 1025 Connecticut Ave, NW, Suite 515, Washington, DC 20036   202-822-6424   1993   32 pages   Free
Ask for this free catalog listing free and inexpensive curriculum materials and classroom activities using plastics.

Local Stores Selling Hands-On Materials and Science Equipment

The following list of retail stores in the Chicagoland area were selected by the author from recommendations by science teachers.

THE ADLER PLANETARIUM SHOP
Adler Planetarium, 1300 S Lake Shore Drive, Chicago, IL 60605   312-922-7827
This store within the Adler Planetarium carries a good selection of books and educational materials on astronomy, astrophysics and space science.

AMERICAN SCIENCE & SURPLUS
5696 Northwest Highway, Chicago, IL 60646   312-763-0313   and 1/4 mile east of Kirk Road, on Route 38, Geneva, IL 60185   708-232-2882   Mail order warehouse: American Science & Surplus, 3605 Howard St, Skokie, IL 60076   708-982-0870
These retail stores are an extensive resource of inexpensive, surplus science

equipment. They are a favorite of science students, science teachers and do-it-yourself inventors. From electric wires and motors to test tubes and telescopes American Science & Surplus seems to have it all. Visit one of these Chicagoland stores and let your creative energy go wild. Ask for a mail order catalog.

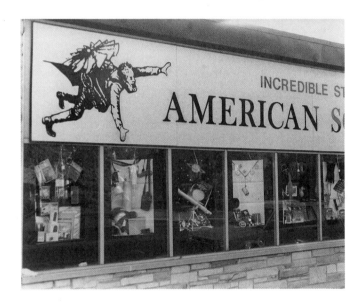

The American Science & Surplus retail stores are a popular source of hands-on materials.

ARVEY PAPER & OFFICE PRODUCTS
(See your Chicago business to business yellow pages under "office supplies" for closest location.) 3555 N Kimball Ave, Chicago, IL 60618 312-463-0822
Discount Office Supplies
This retail store offers an extensive selection of paper and office supplies for that science experiment or classroom need. Every good scientist and inventor knows the value of carefully keeping a log book record of experiments.

BAREBONES
K106, Woodfield Mall, Schaumburg, IL 60173 708-413-2663
Although this store concentrates on fun materials about the human body and its anatomy, you will find fun oddities from geology and optical illusions.

CHESTER ELECTRONICS SUPPLY CO.
1416 Washington St, Waukegan, IL 60085 708-623-5000
Since 1961 this retail store has maintained a large supply of electronic component parts.

CHICAGO BOTANIC GARDEN - THE GARDEN SHOP
Chicago Botanic Garden, P. O. Box 400, 1000 Lake-Cook Road (at Edens Expressway), Glencoe, IL 60022-0400 708-835-8205
This retail shop carries botanical items, items related to wildlife, as well as geological rock specimens.

CHICAGO KITE CO.
6 S Brockway, Palatine, IL 60067 708-359-2556
This retail store specializes in kites. It also has boomerangs and other science toys.

CLASS MATE
3925 W 103rd St, Chicago, IL 60655 312-239-8053
School supplies including science materials for preschool and elementary school.

CONSTRUCTIVE PLAYTHINGS
5314 W Lincoln Ave, Skokie, IL 60076 708-675-5900
This parent/teacher retail store is filled with educational fun for the preschool and elementary school age child.

CUT RATE TOYS
5409 W Devon Ave, Chicago, IL 60646 312-763-5740
For over 40 years this retail store has sold major name toys at discount prices as well as surplus and discontinued items at all price ranges. Science principles and concepts hide in many of these fun toys. A large metal "Slinky" costs $ 1.59.

DAVE'S ROCK SHOP
704 Main St, Evanston, IL 60202 708-866-7374
This lapidary store specializes in minerals, rocks and fossils. Ask to see private museum collection of fossils dating back 1.5 billion years for algae samples and 600 millions years for fossils. Collected over a 35 year period and housed in one room, this collection of Earth's life history includes dinosaur eggs and nests.

DEVON TOYS
2424 W Devon Ave, Chicago, IL 60659 312-973-0194
Former location of Cut Rate Toys. Major name toys at discount prices.

DOOLIN AMUSEMENT SUPPLY CO.
511 N Halsted, Chicago, IL 60622 312-243-9424 Party supplies
A retail source of balloons and novelties for science experiments.

EARTHDWELLER
Two locations: 1734 Sherman Ave, Evanston, IL 60201 708-332-1760
Lincolnwood Town Center (Touhy Ave & McCormick Blvd), Lincolnwood, IL 60645 708-933-0135
These nature and science gift centers carry a large selection of quality educational materials.

EDUCATIONAL TEACHING AIDS
620 Lakeview Station, Vernon Hills, IL 60061 800-445-5985
School supplies including science materials for preschool and elementary school.

GRAY'S DISTRIBUTING CO., INC., - THE LEARNING TREE
4419 Ravenswood Ave, Chicago, IL 60640 312-769-3737
School supplies including science materials for preschool and elementary school. Retail distributor of science materials from the Ideal School Supply Company.

HARRISON SUPPLY
345 N Wolf Rd, 3 blks N of Dundee, Wheeling, IL 708-537-0255
This retail store sells raw materials by the pound, including brass, aluminum and plastic. This is the place for the inventor or science teacher. It also sells motors and much more.

J.C.'S KITES
197 Peterson Rd., Libertyville, IL 60048 708-816-9990
This retail store specializes in kites. It also has boomerangs and other science toys.

KAY BEE TOYS
(See the yellow pages of your telephone directory to locate one of the twenty stores in the Chicagoland area.)
This toy retailer carries toys that use science principles as well as basic science toys.

THE KITE HARBOR
North Pier, 435 E Illinois St, Chicago, IL 60611 312-321-5483
This retail store specializes in kites. It also has science toys like balloon helecopters, boomerangs, the swinging wonder and model planes.

KOHL LEARNING STORE
165 Green Bay Rd, Wilmette, IL 708-251-7168
This retail store within the Kohl Children's Museum is open to the public and sells many quality educational toys for the preschool and elementary school child.

108 Science Fun in Chicagoland

KRASNY & COMPANY INC.
2829 N Clybourn, Chicago, IL 60618 312-477-5504
Restaurant and party supplies. Retail source of clear straws and balloons for science experiments.

LEON M. LEDERMAN SCIENCE EDUCATION CENTER - INSTRUCTIONAL MATERIALS
Fermi National Accelerator Laboratory, P. O. Box 500 MS 777, Batavia, IL 60510 708-840-8258
A variety of educational materials are available for purchase at this center. Cost varies. Contact the center for an order form.

LIGHT WAVE
North Pier, 435 E Illinois St, Chicago, IL 60611 312-321-1123
This retail store specializes in holographic art. Holograms are three dimensional photographs where laser light produces interference patterns on film, where science and art mix to make fun. This store is like a museum with no admission fee.

LIZZADRO MUSEUM OF LAPIDARY ART - SHOP
220 Cottage Hill Ave, Elmhurst, IL 60126 708-833-1616
This excellent museum exhibits art forms made from minerals. It is an example of how art and the science of geology combine to reveal nature's beauty. The shop at the Museum sells many fine examples of minerals, meteorites, petrified wood, and lapidary art.

MIDWEST MODEL
12040 S Aero Dr, Plainfield, IL 60544 815-254-2151
Write for free catalog. Source for Estes Rockets & accessories, bridge building kits, aero-space models, plane programs, mousetrap racers, and model making materials.

MUSEUM OF SCIENCE & INDUSTRY BOOKSTORE & SHOP
Museum of Science & Industry, 57th St and Lake Shore Drive, Chicago, IL 60637 312-684-1414, ext 2780
This retail store has its largest display at the main entrance within the Museum of Science & Industry. This store exclusively sells science books, toys and novelties.

NATURAL WONDERS
Fox Valley Mall, 2330 Fox Valley Center, Aurora 60504   708-820-9004; Stratford Square, 609 Stratford Square, Bloomingdale 60108   708-893-9803; and Oak Brook Center, 504 Oak Brook Ct, Oak Brook 60521   708-954-6613.
Nature and science gift stores. Ask for brochure, *Our Discount for Teachers*, that provides a savings of 15 % on purchases for the classroom.

THE NATURE COMPANY
Watertower Place, 845 N Michigan Ave, Chicago IL 312-751-0330 (Also Northbrook Court, Oakbrook Shopping Center, Wheaton Town Square)
Clothing for hiking, bird feeders and houses, binoculars, electronic pedometer, knives, compasses, Raise and Release Butterfly Kits, tents, kites, Fossil Fish Kit, toy periscope and binoculars. Inquire about teacher discount.

THE ST. JAMES'S HOUSE
4742 W Peterson Ave, Chicago, IL 60646 312-545-0011
This antique shop specializes in clocks, scientific instruments, and maps & books related to the history of science and technology. Sales, repair, restoration, appraisals.

STANTON HOBBY SHOP INC.
4718 N Milwaukee Ave (near Lawrence Ave), Chicago, IL 60630 312-283-6446
This very large hobby store has everything for the young scientist: Estes rockets, dinosaur models, chemistry lab kits, Smithsonian kits, and human anatomy models.

STOREHOUSE OF KNOWLEDGE
2822 N Sheffield, Chicago, IL 60657 312-929-3932
This retail store for school supplies has an extensive section of books and materials on science for the preschool and elementary school age child. As a parent/teacher store, it has many science gift ideas.

SYLVESTER ELECTRICAL SUPPLY CO.
4055 N Kilpatrick Ave, Chicago, IL 60641 312-282-4466
Lighting fixtures, electrical supplies, light bulb center.

THE SCOPE SHOPPE INC.
Box 1208, 113 Read St, Elburn, IL 60119 708-365-9499
Ask for catalog/reference manual. This store sells microscopes for the classroom at all levels of instruction. Educational microscope care, repair and sales.

TOM THUMB HOBBY & CRAFTS
1026 Davis St, Evanston, IL 60201 708-869-9575
This large hobby store has a good selection of hobby materials.

TOY STATION
270 Market Square, Lake Forest, IL 60045 708-234-0180
This store sells many different science kits, Educational Insights Science Kits, microscopes, telescopes and astronomy materials.

TOYS R US
(See the yellow pages of your telephone directory.)
This major toy retailer often carries toys that use science principles.

110 Science Fun in Chicagoland

UNCLE FUN
1338 W Belmont, Chicago, IL 60657   312-477-8223
A unique toy store with hard to find inexpensive toys. They have "hand boilers" that illustrate the science behind the drinking bird and many other toys demonstrating science concepts and processes. The name of this store says it all.

Mail Order Hands-On Materials and Science Equipment Sources

These sources were selected by the author or recommended by science teachers.

ACE-ACME
4100 Forest Park, St. Louis, MO 63108-2899   800-325-7888   Importer, Manufacturer and Distributor   Retail   (All ages)
Ask for 110-page catalog filled with a very large variety of inexpensive toys and novelties that demonstrate natural phenomena. Source for science teachers.

ACTIVITIES INTEGRATING MATHEMATICS AND SCIENCE (AIMS)
AIMS Education Foundation, P. O. Box 8120, Fresno, CA 93747-8120   209-255-6396   Retail   (Elementary School)
Ask for 26-page programs and products catalog. The goal of the AIMS program is to enrich the education of students in K-9 using a hands-on approach that is consistent with *Science for All Americans* (AAAS) and *Curriculum and Evaluation Standards for School Mathematics* (National Council of Teachers of Mathematics).

AMERICAN CHEMICAL SOCIETY
- TEACHING RESOURCES CATALOG
American Chemical Society, Education Division, 1155 16th St, NW, Washington, DC 20036 202-452-2113   Retail   (High School)
Ask for 24-page *Teaching Resources Catalog* that includes posters, videotapes, magazines, educational materials, curriculum supplements, National Chemistry Examinations, and the annual Chemistry Calendar.

ANDY VODA OPTICAL TOYS
P. O. Box 23, Putney, VT   05346   802-387-5457   Manufacturer   Retail   (All ages)
Ask for brochure. Phenakistascope with six magic wheels, Thaumatrope, Couples spinning pictures, Flipbooks, Greeting Flipbooks, Make-it-yourself Zoetrope.

ARBOR SCIENTIFIC
P. O. Box 2750, Ann Arbor, MI 48106-2750 800-367-6695 Retail (High School)
Ask for 35-page catalog of physics demonstration and laboratory equipment.

ARCHIE MC PHEE
P. O. Box 30852, Seattle, WA 98103 206-782-2344 Retail (All ages)
Ask for 30-page catalog. Fun, inexpensive toys and novelties. Insects, bats, turtles, fish, dinosaurs, eye balls, beanie with propeller, iguanas, wall walker octopus. Retail store located at 3510 Stone Way N, Seattle, WA.

CAROLINA BIOLOGICAL SUPPLY CO.
2700 York Rd, Burlington, NC 27215 910-584-0381 Retail (Elementary School to High School)
Major source of science education equipment and supplies.

CENCO
3300 CENCO Parkway, Franklin Park, IL 60131-1364 800-262-3626 Retail (Elementary School to High School)
Major supplier of science education equipment and supplies conveniently located in the Chicago area.

CHILDCRAFT EDUCATION CORP.
20 Kilmer Road, P. O. Box 3081, Edison, NJ 08818-3081 800-631-5652 Distributor Retail (Preschool and Elementary School)
Ask for the 200-page catalog. Math toys, aquariums, plant growing kit, seashells, root garden, ant farm, microscope, optical toys, magnetic toys, weather materials, globes. Educational materials.

COLE-PARMER INSTRUMENT CO.
7425 N Oak Park Ave, Chicago, IL 60648 800-323-4340 Wholesale and Retail (High School to Adult)
Major supplier of scientific equipment and supplies to industry and education and conveniently located in the Chicago area.

CONSTRUCTIVE PLAYTHINGS
1227 E 119th St, Gradview, MO 64030-1117 800-448-4115 Retail (Preschool to Elementary School)
Ask for 184-page catalog filled with educational fun for the preschool and elementary school age child including six pages of hands-on science materials.

112 Science Fun in Chicagoland

CREATIVE PUBLICATIONS
5040 W 111th St, Oak Lawn, IL 60453-5008 800-624-0822 Retail (Elementary School)
Ask for the 115-page catalog filled with educational materials for math, geometry, science, science measurement, and educational curricula. Also science posters.

CUISENAIRE CO. OF AMERICA, INC.
10 Bank St, P. O. Box 5026, White Plains, NY 10602-5026   800-237-0338 Retail (Elementary School)
Ask for 140-page catalog of materials for learning mathematics and science for grades K-9.

DALE SEYMOUR PUBLICATIONS
P. O. Box 10888, Palo Alto, CA 94303-0879 800-872-1100 Retail (Elementary School)
Ask for 160-page catalog of materials and publications for K-8 educational materials in mathematics and science.

DELTA EDUCATION, INC.
P. O. Box 3000, Nashua, NH 03061-3000 800-442-5444 Manufacturer Retail (Elementary School)
Ask for 60-page *Hands-On Science Catalog*. This catalog is filled with science kits, toys and educaional materials. Also ask for information about Delta Science Modules, SCIS3, and ESS curriculum materials. These three hands-on programs are available through Delta Education.

DENOYER-GEPPERT
5225 Ravenswood Ave, Chicago, IL 60640-2028 800-621-1014 Retail (Middle School to High School)
Chicago area distributor of science education equipment, life models, and supplies for biology and chemistry.

DIDAX INC. EDUCATIONAL RESOURCES
395 Main St, Rowley, MA 01969 800-458-0024 Retail (Elementary School)
Ask for 143-page catalog of hands-on materials in elementary school mathematics and science.

DISCOVERY CENTER
Lawrence Hall of Science, University of California, Berkeley, CA   94720 415-642-1016 Retail (All ages)
Ask for a catalog of quality science education hands-on materals.

DOWLING MINER MAGNETICS CORP.
21707 8th St East, P. O. Box 1829, Sonoma, CA 95476   707-935-0352
800-MAGNET-1 Manufacturer Wholesale and Retail (High School to Adult)
Magnets and magnetic materials of numerous shapes and dimensions. Ask for catalog.

EDMUND SCIENTIFIC COMPANY
101 E Gloucester Pike, Barrington, NJ 08007-1380 609-573-6250 Retail (All ages)
Ask for 220-page catalog. Since 1942 this well known scientific optical supplier also sells many other items including lasers, microscopes, camera/monitor systems, science classroom anatomy models, nature kits, laboratory safety equipment, balances, weather instruments, timers, magnets, small motors & pumps, robot kits, earth science kits, telescopes, museum animal replicas, and unique classroom materials for teachers.

These students at Morrill Elementary School in Chicago are enjoying a hands-on science lab developed by their teacher just for them. Their school was one of the first schools to enjoy assistance from the Teachers Academy for Mathematics and Science (TAMS).

114 Science Fun in Chicagoland

EDUCATIONAL TOYS, INC.
P. O. Box 630882, Ojus, FL 33163-0882 800-881-1800 Distributor Retail (Elementary School)
Ask for 30-page catalog. Wild Animal Families models, dinosaur and anatoy puzzles, dinosaur skeleton kits, Farm Life animal models, Monterey Bay Aquarium sea life models, space and mineral materials, The Carnegie Collection of dinosaur models, animal games, endangered animal puzzles, reptile models, Bug Jar, insect posters, toy microscopes and binoculars, Magnetic Marbles, Newton's Cradle swinging balls, kaleidoscope, models of the Rainforest Poison Dart Frog.

ESTES INDUSTRIES
1295 H Street, Penrose, CO 81240 719-372-6565 Retail to teachers. (High School to Adult)
Ask for catalog on school letterhead. Supplies model rockets, engines and accessories.

ETA
620 Lakeview Pkwy, Vernon Hills, IL 60061 800-445-5985 Retail (Elementary School)
Ask for 50-page *ETA Science Catalog* of educational supplies for the science classroom.

EUREKA!
Lawrence Hall of Science, University of California, Berkeley, CA 94720 510-642-1016 Retail (All ages)
Ask for 30-page Lawrence Hall of Science publications catalog, *Eureka!*, filled with books, curriculum materials, science kits and videos.

EXPLORATORIUM STORE
3601 Lyon St, San Francisco, CA 94123 800-359-9899 Retail (All ages)
Ask for mail order catalog. This unique hands-on science museum is well known for its quality science toys and materials.

FIELD MUSEUM OF NATURAL HISTORY - HARRIS EDUCATIONAL LOAN PROGRAM
The Field Museum, Roosevelt Road and Lake Shore Drive, Chicago, IL 60605-2497 312-922-9410, ext 853 Registration fee $ 30, late fee $ 1 per item per day (Elementary School to High School)
Any Chicago-area educator may borrow from an extensive list of Exhibit Cases, Experience Boxes, and Audiovisual Materials. This center is located on the ground floor level of the Museum. Ask for the free 22 page catalog.

FIELD MUSEUM OF NATURAL HISTORY
- THE NEW EXPLORERS MATERIALS DEPOSITORY
The Field Museum, Roosevelt Road at Lake Shore Drive, Chicago, IL 60605-2497   312-922-9410, ext 853 (Elementary School to High School)
The popular PBS television series, *The New Explorers*, is made available to teachers along with educational materials. This collection is found in the Harris Educational Loan Center on the Museum ground floor.

FISHER-EMD - EDUCATIONAL MATERIALS DIVISION
4901 W LeMoyne St, Chicago, IL 60651   General 800-955-1177, College Division 800-955-6644 Retail (High School)
Major supplier of science education equipment and supplies conveniently located in the Chicago area.

FLINN CHEMICAL CATALOG REFERENCE MANUAL
Flinn Scientific Inc., P. O. Box 219, 131 Flinn St, Batavia, IL 60510 800-452-1261   688 pages   Retail   (High School to Adult)
This catalog lists new products, chemicals, chemical solution preparation, Apparatus & Laboratory Equipment, Books, Computer Software, Satety Storage Cabinets & Fume Hoods, Safety Supplies & Equipment, Right to Know Laws, Mystery Substance Identification, Chemical Inventory & Storage, and Chemical Disposal Procedures.

FORESTRY SUPPLIERS, INC.
205 W Rankin St, P. O. Box 8397, Jackson, MS 39284-8397   601-354-3565
Hands-on field, lab and classroom equipment for earth, life and environmental sciences. Orienteering compasses; soil and water sampling, collecting and testing equipment and kits; ecology field instruments, tapes and data collecting instruments; plant and insect collecting materials. Ask for a catalog.

FREY SCIENTIFIC
905 Hickory Ln, P. O. Box 8101, Mansfield, OH 44901-8101   800-225-3739 Retail (Elementary School to High School)
Ask for catalog. Major supplier of hands-on science education equipment and supplies.

THE GEOCENTER (GEOCENTRAL)
1721 Action Ave, Napa, CA 94559 800-231-6083 707-224-7500 Importer and Manufacturer Wholesale and Retail (Elementary School to High School)
Quantity sets of rocks and minerals for retail sale. Flat boxes of mineral and fossil assortments. Sea shell glow night lights. Agate bookends. Twenty page catalog.

IDEA FACTORY, INC.
10710 Dixon Dr, Riverview, FL 33569 800-331-6204 (Elementary and Middle School)
Ask for 16-page catalog filled with books and with science teaching ideas and materials.

IDEAL SCHOOL SUPPLY COMPANY
11000 S Lavergne Ave, Oak Lawn, IL 60453 800-845-8149 Distributor Retail (Preschool to Elementary School)
Ask for the 50-page teacher catalog. Science measurement materials, chemistry experiment beakers and test tubes, equilateral prisms, physics pulleys, themometers, classroom science kits, magnetic toys, natural science materials.

KIPP BROTHERS, INC.
240-242 S Meridian St, P. O. Box 157, Indianapolis, IN 46206 800-428-1153 Importers Retail and Wholesale (All ages)
Ask for 224-page catalog. Established in 1880 this distributor specializes in inexpensive toys, novelties, carnival and party items for quantity, dozen purchases. Science items include dinosaur tattos, animal sounds, musical toys, tops, magnetic wheels, rubber and foam balls, kaleidoscopes, bird gliders, solar radiometer, telescopes, boomerangs, magnetized marbles, flying toys, kazoos, magnifying glasses, museum quality dinosaurs, and many, many more.

KLUTZ
2121 Staunton Court, Palo Alto, CA 94306 415-424-0739 Manufacturer and Distributor Retail (All ages)
Ask for the 70-page *Flying Apparatus Catalogue*. Really fun toys and novelties. Amazon Worms, Smartballs, Smartrings, The Explorabook - a kids science museum in a book, ExploraCenter, Backyard Weather Station kit, Mega-Magnet Set, Backyard Bird Book with bird caller, The Aerobie Orbiter, Rubber Stamp Bug Kit, Vinyl Vermin, Kids Gardening Guide, World Record Paper Airplane Kit, The Arrowcopter, Megaballoons, Bubble Book, Zoetrope, juggling materials.

KOLBE CONCEPTS, INC.
P. O. Box 15667, Phoenix, AZ 85060 602-840-9770 Manufacturer Retail (All ages)
Think-ercises, Glop Shop - inventor's assortment, Go Power - science experiments, Using Your Senses, *Solar Power Winners* - experiment book, *Decide & Design* - inventor's book.

LAB SAFETY SUPPLY INC.
P. O. Box 1368, Janesville, WI 53547-1368 800-356-0783 Retail (High School)
Ask for 900-page catalog of laboratory equipment and supplies dedicated to personal environmental safety for industry, hazardous waste, and school science laboratories.

LAKESHORE LEARNING MATERIALS
2695 E Dominguez St, Carson, CA 90749 800-428-4414 Manufacturer and Distributor Retail (Elementary School)
Ask for 200-page catalog. The catalog of this major distributor of learning materials has eight pages of science materials.

LEARNING RESOURCES, INC.
675 Heathrow Drive, Lincolnshire, IL 60069 708-793-4500 800-222-3909 Manufacturer Wholesale (Preschool and Elementary School)
Educational materials. Mini-Dinos Activity Kit, Story Puzzles with Animals, math and science measurement materials, geometry shapes, wood base ten blocks, thermometers, microscopes, prehistoric animal models, color paddles, Power of Science kits, technology kits, nature kits, measurement sets.

LEGO DACTA
- THE EDUCATIONAL DIVISION OF LEGO SYSTEMS, INC.
555 Taylor Rd, P. O. Box 1600, Enfield, CN 06083-1600 800-527-8339 Manufacturer Retail and Wholesale (Elementary School to High School)
Gear, lever and pulley toys, Technic classroom kits, Technic control centers, teacher's guide books, Pneumatics, Logowriter Robotics for Apple and MS-DOS, Control Lab for Apple and MS-DOS.

LIVING CLASSROOMS
American Forests, P. O. Box 2000, Washington, DC 20013 202-667-3300 Retail (Elementary School to High School)
Living Classrooms includes 20 famous and historic trees plus three years worth of educational curricula.

MATH AND SCIENCE HANDS-ON (M.A.S.H.)
Dr. David A Winnett, Asst Prof, Dept. of Curriculum & Instruction, Southern Illinois University at Edwardsville, P. O. Box 1122, Edwardsville, IL 62026 618-692-6082 Retail (Elementary School)
Ask for brochure listing science kits and materials.

MIDWEST PRODUCTS CO., INC.
400 S Indiana St, P. O. Box 564, Hobart, IN 46342 219-942-1134 Retail (Elementary School to High School)
Source of materials for model aviation and model bridge building.

MORITEX
6440 Lusk Blvd, Suite D-105, San Diego, CA 92121 800-548-7039
This company sells video microscopes that connect to television monitors.

NASCO
901 Janesville Ave, Fort Atkinson, WI 53538-0901 800-558-9595 414-563-2446
Retail (Elementary School to High School)
Ask for 350-page catalog of educational supplies for science.

THE NATURE COMPANY CATALOG
P. O. Box 188, Florence, KY 41022 800-227-1114 Distributor Retail (Elementary School to Adult)
Ask for this 40-page catalog. Clothing for hiking, bird feeders and houses, binoculars, electronic pedometer, knives, compasses, Raise and Release Butterfly Kits, tents, kites, Fossil Fish Kit, toy periscope and binoculars.

NEBRASKA SCIENTIFIC
A Division of Cyrgus Co., Inc., 3823 Leavenworth St, Omaha, NE 68105-1180 800-228-7117 402-346-7214 Retail (Middle School to High School)
Ask for 140-page catalog of educational supplies for the life science classroom including preserved specimens.

OPTICAL SOCIETY OF CHICAGO
640 Pearson St, Suite 200, Des Plaines, IL 60016 708-298-6692 Retail distributor. Single Kits $ 19.95 each. Retail (Elementary School to High School)
Ask for information about Optics Classroom Kits for teachers and students made available by the Optical Society of America.

ORIENTAL TRADING CO., INC.
P. O. Box 3407, Omaha, NE 68103-0407 Orders 800-228-2269, Customer Service 800-228-0475 Distributor Wholesale and Retail (All ages)
Ask for catalog. Source of inexpensive novelties, including magnifying glasses, plastic tops, mini reflectors, balloons, balloon helicopters.

PASCO SCIENTIFIC
10101 Foothills Blvd, P. O. Box 619011, Roseville, CA 95661-9011 800-772-8700 916-786-3800 Retail (High School)
Major supplier of science education equipment and supplies, especially physics.

PENCILS & PLAY
2930 London Rd, Eau Claire, WI 54701 800-659-6154 Retail (Elementary School)
Ask for 280-page catalog that includes 19 pages of science materials for the elementary school classroom.

PLASTICS IN OUR WORLD
American Plastics Council, 1275 K St, NW, Suite 400, Washington, DC 20005 800-243-5790 Free (Elementary School)
Curriculum materials for use with grades K-6 includes How to Set Up a School Recycling Program, Plastics in Perspective, Classroom Activities K-3 and 4-6, and American Plastics Council Materials Order Form.

PLAY VISIONS
1137 N 96th St, Seattle, WA 98103 800-678-8697 206-524-2774 Distributor Wholesale (Elementary School)
Ask for 50-page catalog. Telescopes, optical toys, Giant Rainforest Insects, reptiles & amphibians, Habitat nature model sets, dinosaur toy sets, vinyl Earth balls, Earth Squish Balls, Native American Arrowheads, and many inexpensive novelty items.

THE POWER HOUSE - LEARNING POWER CATALOG
Commonwealth Edison, 100 Shiloh Blvd, Zion, IL 60099 708-746-7080 Retail (Elementary School to High School)
The *Learning Power Catalog* is available to teachers. It lists free materials available, including literature on energy conservation, on the history of electricity, on the environment, on nuclear issues, and on safety as well as videos available for free loan on these same topics.

PROJECT STAR
Learning Technologies, Inc, 59 Walden St, Cambridge, MA 02140 800-537-8703 Retail (Elementary School to High School)
Ask for catalog of economical lab materials, including astronomy labs.

SARGENT-WELCH A VWR COMPANY
911 Commerce Court, Buffalo Grove, IL 60089-2375 800-SARGENT Retail (Elementary School to High School)
Major source of science education equipment and supplies conveniently located in the Chicago area.

SAVI/SELPH
Center for Multisensory Learning, Lawrence hall of Science, University of California, Berkeley, CA 94720 510-642-8941 Retail (Elementary School to Middle School)
Ask for catalog of products including books, curriculum materials, videos and science kits. The Center for Multisensory Learning specializes in science curriculum development for all elementary and junior high school students, including students with disabilities.

120 Science Fun in Chicagoland

SCHOLASTIC SCIENCE PLACE PROGRAM
Contact Marcia Williams, Midwest Representative, Scholastic Inc., 800-225-4625 Retail (Elementary School)
This program offers hands-on experiences for the elementary student that are conveniently packaged in self-contained suitcases that include materials, lessons, and books. Ask for a brochure.

SCIENCE KIT & BOREAL LABORATORES
777 E Park Dr, Tonawanda, NY 14150-6782   800-828-7777   Retail (Elementary School to High School)
Ask for 1000-page catalog. Major supplier of science equipment and science kits for classrooms.

SCOTT RESOURCES
P. O. Box 2121, Fort Collins, CO 80522   303-484-7445 800-289-9299 Distributor Retail (Elementary School)
Ask for 30-page *Earth Science Catalog* and 16 page *Math Materials Catalog*. Earth Science Videolabs, rock and mineral collection trays, large selection of specific rock and mineral samples, classroom projects, fossils, environmental materials, Solar Oven, astronomy materials, solar system simulator, Moon-Earth orbit model, meteorology materials, physical geography models, Earth history educational materials, educational videotapes.

SELSI COMPANY, INC.
40 Veterans Blvd, P. O. Box 497, Carlstadt, NJ 07072   201-935-0388 800-275-7357 Manufacturer Wholesale (Elementary School to Adult)
Quality binoculars, telescopes, student microscope sets, magnifiers, toy kaleidoscopes, glass prisms, student magnets, compasses, barometers, altimeters, metal detectors.

SUMMIT LEARNING
P. O. Box 493, Ft. Collins, CO 80522   800-777-8817 Retail   (Elementary School)
Ask for the 75-page catalog filled with educational materials for math and science including the following categories: Linear Tools; Volume and Capacity; Weights and Measures; Time and Temperature; Problem-Solving; Estimation; Graphs; Probability; Earth Science; Astronomy; and Science and Nature.

THREE RIVERS AMPHIBIAN, INC.
668 Broadway, Dept. GG, Massapequa, NY 11758   516-795-3794   Retail (Elementary School)
Ask for brochure including mail order living Growa-Frog Kit, Frog Friend, Tadpoles, and literature, *Pollywogs 'n Frogs*.

Chapter 09 - Materials

TIMS (TEACHING INTEGRATED MATHEMATICS AND SCIENCE)
TIMS Project, (M/C 250), Institute for Mathematics and Science Education, University of Illiois at Chicago, P. O. Box 4348, Chicago, IL 60607 312-996-2448 TIMS Equipment and Supplies: Hubbard Scientific, 1120 Halbleib Rd, P. O. Box 760, Chippewa Falls, WI 54729 800-323-8368 715-723-4427 Retail (Elementary School)
Ask the Institute for the *Documents Catalog* that describes available experiments and the TIMS Tutors for teachers. TIMS, in use since 1974, is a quantitative hands-on approach to K-8 science that uses fun scientific experimental methods and thinking to integrate math and science. Contact the Institute for information about availability of illustrated children's story books being developed for TIMS experiments. Ask about the *Extrapolator*, newsletter published by the Institute. Individual TIMS experiments and Tutors are also available to interested parents.

WARD'S NATURAL SCIENCE ESTABLISHMENT, INC.
5100 W Henrietta Rd, P. O. Box 92912, Rochester, NY 14692-9012 800-962-2660 Retail (Elementary School to High School)
Major supplier of science education equipment and supplies.

WISCONSIN FAST PLANTS
University of Wisconsin-Madison, Dept. of Plant Pathology, 1630 Linden Dr, Madison, WI 53706 800-462-7417 Retail (Elementary School to High School)
This source of Brassica Rapa, a plant that takes 34 days growing from seed to seed. This plant is useful in genetic studies. Also, information on bottle biology for the classroom.

YOUNG ENTOMOLOGIST'S SOCIETY, INC.
1915 Peggy Place, Lansing, MI 48910-2553 517-887-0499 Retail (Elementary School to High School)
Ask for the 70-page *Buggy Bookstore Catalog* including programs and services, products and materials, and tips for selecting materials. Resource materials include books, educational games, models, audiovisual materials, and toys. Inquire about membership.

Chapter 10
Periodicals

Periodical Directories

GENERAL SCIENCE INDEX
H. W. Wilson Co., 950 University Ave, Bronx, NY 10452-4297   212-588-8400
Monthly Bound annually. Available in most major libraries.
Cumulative index to published works in English language periodicals. Articles listed by subject and author with quarterly and annual cumulations published.

THE STANDARD PERIODICAL DIRECTORY
Oxbridge Communications, Inc., 150 Fifth Ave, Suite 302, New York, NY 10011 Published annually. Available in most major libraries.
This guide to United States and Canadian periodicals alphabetically lists over 85,000 pubications under subject categories.

Chapter 10 - Periodicals 123

ULRICH'S INTERNATIONAL PERIODICALS DIRECTORY
R. R. Bowker, A Reed Reference Publishing Company, 121 Chanion Rd, New Providence NJ 07974 800-346-6049 Published annually Available in most major libraries.
Published in four volumes, periodicals are alphabetically listed by category.

Periodicals for Parents and Children

Perhaps one of the nicest gifts a child can receive is a magazine subscription addressed to the child's name.

3-2-1 CONTACT
Children's Television Workshop, P. O. Box 51177, Boulder, CO 80322-1177 Published 10 times per year. $ 16.97 per year.
A science and technology magazine for young children based on the television series of the same name.

ABRAMS PLANETARIUM SKY CALENDAR
Abrams Planetarium, Michigan State University, East Lansing, MI 48824 Quarterly $ 6.00 per year.
For the beginning astronomer, this calendar presents a nightly description of important astronomical events and a simplified chart of the evening sky.

AMERICAN KITE
American Kite Company, 13355 Grass Valley Ave, Grass Valley, CA 95945 916-273-3855 Quarterly $ 14.00 per year.
This quarterly journal is dedicated to kiting.

APPRAISAL: SCIENCE BOOKS FOR YOUNG PEOPLE
Boston University School of Education, 605 Commonwealth Ave, Boston, MA 02215 617-353-4150 Quarterly $ 42.00 per year.
Book reviews of new science books for children are presented by both a librarian and a science specialist.

BABYBUG
Box 723, Mt. Morris, IL 61054-8112 800-827-0227 Every 6 weeks. $ 29.97 per year for 9 issues.
Babybug is a new listening and looking magazine for infants and toddlers ages 6 months to 2 years. This board-book 24 page magazine has rounded corners and is just the right size for small hands.

CHICKADEE
The Young Naturalist Foundation, 255 Great Arrow Ave, Buffalo, NY 14207-3082; The Young Naturalist Foundation, 56 The Esplanade, Suite 306, Toronto, Ontario M5E 1A7, Canada   416-868-6001   Published 10 times each year. $ 21.00 per year.
Children learn about environment, nature and science.

CURRENT SCIENCE
Weekly Reader Corp., 3001 Cindel Dr, Delran, NJ 08370   800-446-3355   Bimonthly $ 7.45 per year.
This newsmagazine presents topics in science, health and technology for grades 7-9.

THE DOLPHIN LOG
The Cousteau Society, 870 Greenbrier Circle, Suite 402, Chesapeake, VA 23320   804-523-9335   Bimonthly $ 10.00 per year.
Dedicated to oceanography and the environment for ages 7-15.

KITELINES
Aeolus Press, Inc., 8807 Liberty Road, Randallstown, MD 21133   410-922-1212   Quarterly $ 14.00 per year.
This comprehensive international journal of kiting is endorsed by the International Kitefliers Association.

NATIONAL GEOGRAPHIC WORLD
National Geographic Society, 17th and M St, NW, Washington, DC 20036   202-857-7296   Monthly $ 10.95 per year.
A natural history magazine of general interest to children.

ODYSSEY
Cobblestone Publishing, Inc., 7 School St, Peterborough, NH 03458-1470   603-924-7209   Published 10 times each year. $ 22.95 per year.
Magazine of astronomy and outerspace written for readers ages 8 to 14.

OWL
Young Naturalist Foundation, 255 Great Arrow Ave, Buffalo, NY 14207; The Young Naturalist Foundation, 56 The Esplanade, Suite 306, Toronto, Ontario M5E 1A7, Canada   416-868-6001   Published 10 times each year. $ 21.00 per year.
A magazine for children ages 8-14 interested in science and nature.

QUANTUM
Springer-Verlag New York, P. O. Box 2485, Secaucus, NJ 07096 800-SPRINGER 201-348-4033 Bimonthly $ 18.00 per year. (800-722-NSTA $ 20 per year)
Dedicated to state-of-the-art math and science for students (and their teachers), this magazine is also available with NSTA membership. Quantum is the English-language version of the Russian journal Kvant.

RANGER RICK'S NATURE MAGAZINE
National Wildlife Federation, 8925 Leesburg Pike, Vienna, VA 22184-0001 703-790-4000 Monthly $ 15 per year.
A nature, environment, and outdoors magazine for children ages 6-12.

SCIENCE NEWS: THE WEEKLY NEWSMAGAZINE OF SCIENCE
Science News, P. O. Box 1925, Marion, OH 43305 800-347-6969 Weekly $ 39.50 per year.
About new discoveries in science and the people who made these discoveries.

SCIENCE WEEKLY
Science Weekly, Inc., 2141 Industrial Parkway, Suite 202, Silver Spring, MD 20904 301-680-8804 Weekly $ 8.95 per year. $ 3.95 per year for 20 or more subscriptions.
Current and high interest science and technology topics for grades K-8.

SCIENCE WORLD
Scholastic Inc., Subscriptions, 2931 E McCarty St, P. O. Box 3710, Jefferson City, MO 65102-9957 800-631-1586 Biweekly $ 7.25 per school year for orders of 10 or more.
For grades 7-10. Science news in general science, nature study, earth science, life and space science.

SCIENCELAND
Scienceland Inc., 501 Fifth Ave, Suite 2108, New York, NY 10017-6102 212-490-2180 Published 8 times each year. $ 19.95 per year.
Picture book early reader for preschool children through primary grades, ages 5-10. Teacher's manual also available.

SUPERSCIENCE BLUE EDITION
Scholastic Inc., Subscriptions, 2931 E McCarty St, P. O. Box 3710, Jefferson City, MO 65101-3710 800-631-1586 Monthly during school year. $ 6.75 per year for 10 or more subscriptions.
A classroom magazine for students in grades 4-6, including science classroom activities.

126 Science Fun in Chicagoland

SUPERSCIENCE RED EDITION
Scholastic Inc., 2931 E McCarty St, P. O. Box 3710, Jefferson City, MO 65101-3710   800-631-1586   Monthly during school year. $ 5.95 per year for 10 or more subscriptions.
A classroom magazine for students in grades 1-3, including science classroom activities.

THE DINOSAUR TIMES
CSK Publications, 299 Market St, Saddlebrook, NJ 07662   201-712-9300   Bimonthly $ 2.50 per issue.
Dedicated to factual articles about dinosaurs as well as myths about dragons and monsters, this publication presents the difference between science and fantasy.

WEATHERWISE
Heldref Publications, 1319 18th St, NW, Washington, DC 20036   800-365-9753   Bimonthly $ 30.00 per year.
Articles about the weather and its relationship to people presented with color photography.

WONDERSCIENCE
American Chemical Society, Education Division, ACS Career Education, Room 806, 1155 16th St, NW, Washington, DC 20036 202-452-2113 Individual back issues are available.
This magazine is in colorful comic book format for grades 4-6. For home or school, these magazines relate science to technology and are also available in bilingual (Spanish/English) editions. Telephone to inquire about availability.

YOUR BIG BACKYARD
National Wildlife Federation, 8925 Leesburg Pike, Vienna, VA 22184   703-790-4000 Monthly $ 12.00 per year.
Simple text and photos about animals, nature and conservation help young children, ages 3-5, learn about science.

Periodicals for Teachers

Contact local libraries to see most of these periodicals before subscribing or ask the publisher for a complimentary examination copy.

ACCELERATOR NEWSLETTER
Saskatchewan Science Teachers Society, Saskatchewan Teacher's Federation, Box 1108, 2317 Arlington Ave, Saskatoon, SK S7J 2H8 CANADA   Since 1964.   Published 4-5 times per year. $ 20.00 per two years.
Newsletter for science teachers.

AMERICAN JOURNAL OF PHYSICS
American Association of Physics Teachers, One Physics Ellipse, College Park, MD 20740-3845   301-209-3300   Monthly $ 224.00 per year.
About physics and teaching physics at the college level.

THE AMERICAN BIOLOGY TEACHER
National Association of Biology Teachers, 11250 Roger Bacon Dr, #19, Reston, VA 22090-5202   703-471-1134   Published 8 times per year. $ 50 per year.
This journal includes specific how-to-do-it suggestions for the laboratory, field activities, programs, and review on recent advances in life science.

CAROLINA TIPS
Carolina Biological Supply Co., 2700 York Rd, Burlington, NC 27215-3398   919-584-0381   Free
Ask to be placed on mailing list. Includes articles that present new teaching materals and tips.

CHEM MATTERS
American Chemical Society, Education Division, ACS Career Education, Room 806, 1155 16th St, NW, Washington, DC   20036   202-452-2113   Published 4 times per year. $ 3.75 per subscription.
This magazine is for high school chemistry students.

CHEMICAL EDUCATION
The Division of Chemical Education of the American Chemical Society Inc., Subscription and Book Order Department, 1991 Northampton St, Easton, PA 18042   202-872-4600   Monthly $ 32 per year.
This journal is dedicated to publishing articles about or relevant to the teaching of chemistry.

CHEMUNITY NEWS
American Chemical Society, Education Division, 1155 16th St, NW, Washington, DC 20036   Monthly   Free
A publication of the ACS Education Division focusing on prehigh school science, high school chemistry, college chemistry, and continuing education.

A CLASS ACT
Argonne National Laboratory, Division of Educational Programs, 9700 S Cass Ave, Argonne, IL 60439   Free
This newsletter is an Argonne Community of Teachers (ACT) publication that includes news items for science teachers. ACT's purpose is one of linking research and education.

COMPRESSED AIR MAGAZINE
253 E Washington Ave, Washington, NJ 07882-2495 908-850-7817 Published 8 times per year. Free
A magazine with interesting articles on applied technology and industrial management. Published by a division of Ingersoll-Rand. Fun reading for teachers, students and those interested in science. Request a subscription application.

THE COMPUTING TEACHER
International Society for Technology Education, 1787 Agate St, Eugene, OR 97403-1923 503-346-4414
This professional organization is dedicated to the improvement of all levels of education through the use of computer-based technology.

CURRENT SCIENCE
245 Long Hill Road, Middletown, CT 06457 800-248-1882 Biweekly $ 5.50 per year for 10 or more sent to the same school address.
For students in science classes in middle and senior high schools.

ELECTRONIC LEARNING
Available from the International Society for Technology Education, 1787 Agate St, Eugene, OR 97403-1923 503-346-4414
This professional organization is dedicated to the improvement of all levels of education through the use of computer-based technology.

EXTRAPOLATOR
Newsletter of the Institute for Mathematics and Science Education, Room 2075 Science & Education Laboratories, University of Illinois at Chicago, 840 W Taylor St, Chicago, IL 60607 312-996-2448 Free
This newsletter is sent to over 2000 science education professionals and describes the curriculum materials and programs available at the Institute for Mathematics and Science Education, including TIMS (Teaching Integrated Mathematics and Science Project).

JOURNAL OF GEOLOGICAL EDUCATION
National Association of Geology Teachers, Inc., P. O. Box 5443, Bellingham, WA 98227-5443 206-650-3587 Published 5 times each year. $ 25 per year includes membership.
This journal seeks to foster improvement in teaching earth sciences at all levels.

GEOTIMES
American Geological Institute, Communications Department, 4220 King St, Alexandria, VA 22302 703-379-2480 Monthly $ 24.95 per year.
This journal reports events, research, meetings, and developments in geoscience education, political activities, and technological advances.

Chapter 10 - Periodicals 129

JOURNAL OF COLLEGE SCIENCE TEACHING
National Science Teachers Association, 1840 Wilson Blvd, Arlington, VA 22201-3000 800-722-NSTA Monthly
Dedicated to college introductory science teaching, this journal is available with NSTA membership. It includes ideas for classroom teachers.

JOURNAL OF RESEARCH IN SCIENCE TEACHING
Journal for the National Association for Research in Science Teaching John Wiley & Sons, Inc., Journals, 605 Third Ave, New York, NY 10158-0012 212-850-6000 Published 10 times per year. $ 246.00 per year.
Articles on research related to the teaching of science.

MUSEUM NOTEBOOK
Education Department, Museum of Science & Industry, 57th St and Lake Shore Drive, Chicago, IL 60637 312-684-1414, ext 2429 Published three times a year. Free
Teachers, ask to be put on the mailing list of this newsletter for educators. It covers upcoming programs, exhibits, and events at the Museum. mailing list.

NSRC NEWSLETTER
National Science Resources Center (NSRC), Smithsonian Institution, Arts & Industries Building, Room 1201, Washington, DC 20560 202-357-2555 Free
Ask to be placed on the mailing list for the NSRC Newsletter. NSRC disseminates information about effective science teaching, develops curriculum materials, and sponsors outreach and leadership development activities.

NSTA REPORTS!
National Science Teachers Association, 1840 Wilson Blvd, Arlington, VA 22201 Bimonthly Included with NSTA membership.
This newspaper format bulletin contains over 50 pages of news items relevant to science teaching.

NSTA'S EDUCATIONAL HORIZONS
Elsie Diven Weigel, NSTA's Educational Horizons, Code FE, 300 E St, SW, Washington, DC 20546-0001 202-358-1533 Quarterly Free
Ask to be placed on the mailing list of this NASA educational newsletter that includes current NASA news, educational division activites, and lists of educational materials.

OREGON SCIENCE TEACHER
Oregon Science Teachers Association, 700 Pringle Pkwy SE, Salem, OR 97310-0001 503-667-5489 Monthly $ 8.00 per year.
Of interest to science teachers, kindergarten to 12th grade.

THE PHYSICS TEACHER
American Association of Physics Teachers, One Physics Ellipse, College Park, MD 20740-3845   301-209-3300   Monthly
About teaching introductory physics at the high school and college level.

THE POWER HOUSE - GAZETTE
Commonwealth Edison, 100 Shiloh Blvd, Zion, IL 60099   708-746-7080
Quarterly newsletter   Free
This newsletter focuses on the topic of energy and energy education related news. It is available to the general public.

QUANTUM
National Science Teachers Association, 1840 Wilson Blvd, Arlington, VA 22201-3000   800-722-NSTA   Six issues per year. $ 20 per year.
Dedicated to state-of-the-art math and science for students (and their teachers), this magazine is also available with NSTA membership.

SCHOOL SCIENCE AND MATHEMATICS
School Science and Mathematics Association, 126 Life Science Building, Bowling Green State University, Bowling Green, OH 43403-0256   419-372-7393   Monthly
Founded in 1901, the purpose of this association is to disseminate research findings and its implications for school practice. Topics in science and mathematics education at the elementary and high school levels.

SCIENCE ACTIVITIES
Heldref Publications, Inc., 1319 18th St, NW, Washington, DC 20036-1802   202-296-6267   Quarterly $ 30.00 per year.
Articles on classroom science projects for teachers of science.

SCIENCE AND CHILDREN
National Science Teachers Association, 1840 Wilson Blvd, Arlington, VA 22201-3000   800-722-NSTA   Monthly
Dedicated to preschool through middle school science teaching, this journal is available with NSTA membership. It includes ideas for classroom teachers.

SCIENCE EDUCATION
John Wiley & Sons, Inc., Journals, 605 Third Ave, New York, NY 10158-0012   212-692-6000   Bimonthly $ 205.00 per year.
This journal reports research on practices, issues and trends in science instruction as well as on the preparation of science teachers for the science classroom and laboratory.

SCIENCE EDUCATION NEWS
American Association for the Advancement of Science, 1333 H St, NW,

Washington, DC 20005-4707 202-326-6620 Monthly 4 pages.
This newsletter reports on current science, mathematics and technology news items for schools.

SCIENCE SCOPE
National Science Teachers Association, 1840 Wilson Blvd, Arlington, VA 22201-3000 800-722-NSTA Monthly
Dedicated to middle school science teaching, this journal is available with NSTA membership. It includes ideas for classroom teachers.

JOURNAL OF SCIENCE TEACHER EDUCATION
Association for the Education of Teachers in Science, Dr. Joe Peters, University of West Florida, 11000 University Pkwy, Pensacola, FL 32514-5753 Quarterly
This journal serves as a forum for presentation and discussion of issues relating to professional development in science teaching.

The NASA Teacher Resource Center at the Museum of Science & Industry is open to teachers by appointment. It contains numerous educational books, slides, video tapes, and posters about the space program.

132 Science Fun in Chicagoland

THE SCIENCE TEACHER
National Science Teachers Association, 1840 Wilson Blvd, Arlington, VA 22201-3000 800-722-NSTA Monthly
Dedicated to middle and high school science teaching, this journal is available with NSTA membership. It includes ideas for classroom teachers.

SCIENCELINES
Teacher Resource Center, Fermi National Accelerator Laboratory, Leon M. Lederman Science Education Center, P. O. Box 500 MS 777, Batavia, IL 60510 708-840-8258 Quarterly newsletter Free
This newsletter contains current information about science education. resources, reviews, activities and scientist interviews as well as articles about Fermilab.

SPECTRUM
- JOURNAL OF THE ILLINOIS SCIENCE TEACHERS ASSOCIATION
Diana Dummitt, Associate Editor ISTA SPECTRUM, University of Illinois, College of Education, 1310 S Sixth St, Champaign, IL 61820 217-244-0173 Quarterly $ 20.00 regular membership includes SPECTRUM.
This journal includes news of ISTA activities, articles on science and science education, and items of interest to science teachers K-12.

SSMART NEWSLETTER
School Science and Mathematics Association, 126 Life Science Building, Bowling Green State University, Bowling Green, OH 43403 419-372-7393 Quarterly
Founded in 1901, the purpose of this association is to disseminate research findings and its implications for school practice.

TEACHING SCIENCE
Association for Science Education, College Lane, Hatfield, Hertfordshire AL10 9AA ENGLAND Published 3 times per year. Six pounds per year.
Articles and ideas of interest for science and technology teachers for grades K-8.

TECHNOLOGY AND LEARNING MAGAZINE
Available from the International Society for Technology Education, 1787 Agate St, Eugene, OR 97403-1923 503-346-4414 Published by Peter Li Education Group, Dayton, OH 45439-1597 513-847-5900 8 issues per year. $ 19.95 per year.
The professional organization, ISTE, is dedicated to the improvement of all levels of education through the use of computer-based technology. This magazine is about technology in the classroom.

THE TECHNOLOGY TEACHER
International Technology Education Association, 1914 Association Dr, Reston, VA 22091 703-860-2100 Monthly $ 55.00 per year.

Each issue provides ideas for the classroom, project activities, resources in technology, and current trends in technology education.

Science Periodicals

Periodicals exist for almost every science topic. The periodicals listed here may be found in your local public library:

AIR & SPACE /SMITHSONIAN
Smithsonian Institution, 900 Jefferson Dr, Washington, DC 20560 800-766-2149 Bimonthly $ 18.00 per year.
Includes articles on aviation, aviation history and space travel.

AMERICAN SCIENTIST
Sigma Xi, Scientific Research Society, Box 13975, 99 Alexander Dr, Research Triangle Park, NC 27709 919-549-0097 Bimonthly $ 28.00 per year.
Sigma Xi is an honor society for scientists and engineers. Articles concentrate on all fields of research in science and technology.

ARCHAEOLOGY
Archaeological Institute of America, 135 Williams St, New York, NY 10038 800-829-5122 Bimonthly $ 19.97 per volume.
Articles include topics on archaeological discoveries and relevant issues.

ASTRONOMY MAGAZINE
Kalmbach Publishing Co., 21027 Crossroads Circle, P. O. Box 1612, Waukesha, WI 53187 800-446-5489 Monthly $ 27.00 per year.
Covers all aspects of astronomy. Articles include reviews such as "Best Astronomy Books for Kids."

AUDUBON
National Audubon Society, 700 Broadway, New York, NY 10003 212-979-3000 Bimonthly $ 20.00 per year.
This magazine of the the National Audubon Society covers wildlife, wilderness and environmental topics.

BIOSCIENCE
American Institute of Biological Sciences, 730 11th St, NW, Washington, DC 20001-4521 202-628-1500 Monthly $ 52.00 per year.
Articles on current research for professional biologists.

BROOKFIELD ZOO PREVIEW
Brookfield Zoo/Chicago Zoological Society, 3300 S Golf Rd, Brookfield, IL

134 Science Fun in Chicagoland

60513 708-485-0263, ext 361 Quarterly Free to members.
The *Preview* includes news about Brookfield Zoo, its exhibits, and its programs for children and the public.

CD-ROM WORLD
Meckler Corporation, 11 Ferry Lane West, Westport, CT 06880-5880 203-226-6967 Monthly $ 29.00 per year.
Information on CD-ROM products for home and professional use.

CENCO REFLECTIONS
News and Notes of Interest, CENCO, 3300 CENCO Parkway, Franklin Park, IL 60131 800-262-3626 Free
Ask to be put on mailing list of this bulletin of ideas using CENCO equipment.

COMPUTE
Compute Publications, International Ltd., 1965 Broadway, New York, NY 10023-5965 800-727-6937 Monthly $ 19.95 per year.
Each issue contains feature articles, columns, multimedia articles, reviews, and an entertainment section.

COSMIC QUARTERLY
Chicago Astronomical Society, P. O. Box 30287, Chicago, IL 60630-0287 Quarterly
This newsletter is included with membership in the Chicago Astronomical Society.

DEEP SKY
Kalmbach Publishing Co., 21027 Crossroads Circle, Waukesha, WI 53187 800-446-5489 Quarterly $ 12.00 per year.
For deep-sky observers and astrophotographers.

DISCOVER - DISCOVERY CENTER MUSEUM
711 N Main St, Rockford, IL 61103 815-963-6769 Quarterly
Newsletter of the Discovery Center Museum containing Museum news, special events, and science topics. The Museum has over 100 hands-on science exhibits inside the museum and even more in the outdoor science park.

DISCOVER
114 Fifth Ave, New York, NY 10011 800-829-9232 Monthly $ 29.95 per year.
Science topics in an illustrated magazine format for the general public.

FACTS ON FILE SCIENTIFIC YEARBOOK
Facts on File, Inc., 460 Park Ave S, New York, NY 10016-7382 212-683-2244 Published annually. $ 30.00 per year.
The year's scientific achievements and developments described for high school students.

FUTURES - EVANSTON ENVIRONMENTAL ASSOCIATION
Evanston Ecology Center, 2024 McCormick Blvd, Evanston, IL 60201 708-864-5181 Quarterly
Newsletter of the Evanston Environmental Association that describes current programs and events at the Center.

GARDEN TALK - CHICAGO BOTANIC GARDEN
Chicago Horticultural Society, P. O. Box 400, 1000 Lake-Cook Road (at Edens Expressway), Glencoe, IL 60022 708-835-5440 Monthly
This newletter of the Chicago Horticultural Society includes Chicago Botanic Garden news, events calendar, and gardening topics.

GRAND VALLEY STATE UNIVERSITY - SCIENCE & MATHEMATICS UPDATE
Grand Valley State University, Science and Math Center, 1 Campus Dr, Allendale, MI 49401-9986 Quarterly Free
Ask to be placed on mailing list of this bulletin on science teaching ideas. Sent to school addresses only.

HORTICULTURE
Horticulture, Inc., 98 N Washington St, Boston, MA 02114-1913 617-742-5600 Published 10 times per year. $ 26.00 per year.
The magazine of American gardening.

I.C.E. CUBE
Computer Update Bulletin for Educators (CUBE), Illinois Cumputing Educators (I.C.E.), 8548 145th St, Orland Park, IL 60462-2839 Contact Vicki Logan at 708-894-8574 Bimonthly $ 25.00 per year, includes membership.
This organization focuses on utilizing computer technology in the classroom. This newletter includes information about computer bulletin boards, information on grants, reviews of software, announcements about meeting where public domain software is traded.

ILLINOIS STATE ACADEMY OF SCIENCE TRANSACTIONS
Illinois State Museum, Spring & Edwards Sts, Springfield, IL 62706 217-782-6436
Articles about new developments in science written by members of the academy.

ISGS GEONEWS
Illinois State Geological Survey, 615 E Peabody Dr, Champaign, IL 61820-6964 217-333-4747 Quarterly
Newsletter of the Illinois State Geological Survey. Includes articles and lists publications available.

136 Science Fun in Chicagoland

ISIS
University of Chicago Press, Journals Division, 5720 S Woodlawn Ave, Chicago, IL 60637-1603   312-753-3347   Quarterly $ 49.00 per year.
Covers the history of science and its cultural influence.

JOURNAL OF THE INSTITUTE OF ENVIRONMENTAL SCIENCES
Institute of Environmental Sciences, 940 E Northwest Hwy, Mt. Prospect, IL 60056   708-255-1561   Bimonthly $ 35 per year.
Areas of interest in the Journal pertain to environmental sciences, product design, and contamination control.

LIFE SCIENCES
Elsevier Science, 660 White Plains Rd, Tarrytown, NY 10591-5153   914-524-9200   Weekly $ 2,140 per year.
International scholarly publication on new work in the bio-medical sciences.

MAC-USER
MacUser, 950 Tower Lane, 18th Floor, Foster City, CA 94404   415-378-5600   800-627-2247   Monthly   Single copy $ 2.95. $ 27.00 per year.
This 200-page magazine is dedicated to the Apple computer and includes news, reviews, and advertisements.   Available where magazines are sold.

MACWORLD
Macworld Communications, 501 Second St, San Francisco, CA 94107   Editorial, 415-243-0505   Subscriptions, 800-288-6848   Monthly   Single issue $ 3.95. $ 30.00 per year.   Available where magazines are sold.
This 250-page magazine is dedicated to Apple computers.   Macworld is a publication of Macworld Communications and is an independent journal not affiliated with Apple Computer, Inc.

MCGRAW-HILL YEARBOOK OF SCIENCE AND TECHNOLOGY
McGraw-Hill, 1221 Ave of the Americas, New York, NY 10020-1095   212-512-4653   Published annually.
An annual summary of achievements in science and technology.

THE MORTON ARBORETUM - EVENTS, NEWS & CLASSES
The Morton Arboretum, Route 53 (just north of interstate 88), Lisle, IL 60532   708-719-2400   Quarterly
A 27-page newletter listing events, news and classes available.

THE MORTON ARBORETUM QUARTERLY
The Morton Arboretum, Route 53 (just north of interstate 88), Lisle, IL 60532   708-719-2400   Quarterly
A 30-page journal with articles about the horticulture of woody plants.

NATURAL ENQUIRER
Spring Valley Nature Sanctuary, 1111 E Schaumburg Rd, Schaumburg, IL 60194 708-980-2100 Bimonthly
Newsletter of the Spring Valley Nature Sanctuary with science topics, news, and events at the Sanctuary.

NATURAL HISTORY
American Museum of Natural History, Central Park West at 79th St, New York, NY 10024-5192 212-769-5500 Monthly $ 28.00 per year.
Articles written by scientists on social and natural science.

NATURE
Nature, Box 1733, Riverton, NJ 08077-7333 800-524-0384 Weekly $ 395.00 per year.
Articles about new discoveries and research in all fields of science.

NATURE'S NOTES - CHICAGO ACADEMY OF SCIENCES
2001 N Clark St, Chicago, IL 60614 312-549-0606, ext 2057 Quarterly
Newsletter on exhibits, field trips, programs and special events.

NEW SCIENTIST
Magazines, Specialist Magazine Group, King's Reach Tower, Stamford St, London SE1 9LS ENGLAND 071-261-7301 Weekly $ 130.00 per year.
Comprehensive coverage of new science research and discoveries in all fields of science.

OMNI
Omni International, Ltd., 324 W Wendover Ave, Suite 205, Greensboro, NC 27408 919-275-9809 Monthly $ 24.00 per year.
Covers many areas of science including medicine, technology, physics and space exploration as well as science fiction.

PC MAGAZINE:
THE INDEPENDENT GUIDE TO PERSONAL COMPUTING
Ziff-Davis Publishing Company, L.P., One Park Ave, New York, NY 10016-5802 Bimonthly Single copy $ 3.95. $ 49.97 per year.
This 400-page magazine is dedicated to all aspects of personal computing. Available where magazines are sold.

PC WORLD
PC World Communications, Inc., 501 Second St, # 600, San Francisco, CA 94107 415-243-0500 Bimonthly Single issue $ 3.95. $ 29.90 per year.
This 350-page magazine is dedicated to news about new products for home computer needs and includes numerous reviews and resources. Available where magazines are sold.

PHILOSOPHY OF SCIENCE
Philosophy of Science Association, Michigan State University, Department of Philosophy, 503 S Kedzie Hall, East Lansing, MI 48824-1032   517-353-9392   Quarterly $ 60.00 per year.
Philosophical analysis of concepts or words used in science.

PHYSICS TODAY
American Institute of Physics, American Center for Physics, One Physics Ellipse, College Park, MD 20740-3843   301-209-3040 Monthly $ 35.00 per year with affiliated society membership.
News about current physics research or research related to physics as well as topics on physics of interest to the general reader.

POPULAR SCIENCE
Times-Mirror Co., 2 Park Ave, New York, NY 10016-5675   212-779-5000   800-289-9399   Monthly $ 13.94 per year.
Written for the general reader, this magazine describes new products and ideas from science and technology.

RE-ACTIONS
555 N Kensington Ave, La Grange Park, IL 60525 800-323-3044   Published five times each year. Free
Published by the American Nuclear Society, this bulletin is for educators interested in learning and teaching about various peaceful uses of nuclear science and careers in the field.

RESOURCE - SCIENCE KIT & BOREAL LABORATORES
Laura Glass, Editor Science Kit & Boreal Laboratories, 777 E Park Dr, Tonawanda, NY 14150-6784 800-828-7777   Free
Ask to be placed on the mailing list of this publication that includes classroom tested science activities.

SCIENCE
American Association for the Advancement of Science, 1333 H St, NW, Washington, DC 20005-4707 202-326-6500 Weekly $ 92.00 per year, includes AAAS membership.
This prestigous weekly journal of science contains articles on original research and science news.

SCIENCE BOOKS & FILMS
by the American Association for the Advancement of Science, 1333 H St, NW, Washington, DC 20005 202-326-6500 Nine issues per year. $ 40.00 per year.
This periodical reviews scientific accuracy and presentation of print, audiovisual, and electronic resources intended for educational use.

SCIENCE ASSOCIATION FOR PERSONS WITH DISABILITIES GOOD NEWSLETTER
edited by Dr. John Stiles, Department of Curriculum and Instruction, University of Northern Iowa, Cedar Falls, IA 50614 Contact Dr. Janet Davies, President, Science Association for Persons with Disabilties, P. O. Box 17441, Boulder, CO 80308-0411 303-666-9312 Newsletter $ 10.00 per year.
Newsletter published by Science Association for Persons with Disabilities.

SCIENCE FOR THE PEOPLE
Science Resource Center, Inc., Box 364, Somerville, MA 02143-0005 617-547-3580 Bimonthly $ 15.00 per year.
About how science and technology can best meet human needs, including topics about nutrition, health, nuclear technology, and workplace safety.

SCIENCE ILLUSTRATED
8428 Holly Leaf Dr, McLean, VA 22102 703-356-1688 Bimonthly $ 15.00 per year.
Popular reading about science news.

SCIENCE NEWS
Science Service, Inc., 1719 N St, NW, Washington, DC 20036-2888 202-785-2255 Weekly $ 44.50 per year.
Overview of science news in all fields of science.

THE SCIENCES
New York Academy of Sciences, 2 E 63rd St, New York, NY 10021-7210 212-838-0230 Bimonthly $ 18.00 per year.
This periodical is written by scientists for both the scientist and the non-scientist.

SCIENTIFIC AMERICAN
Scientific American, Inc., 415 Madison Ave, New York, NY 10017-1179 212-754-0550 Monthly $ 36.00 per year.
This magazine is about a broad range of science topics presented at a technical level for persons in professional positions.

THE SCIENTIST
Scientist, Inc., 3600 Market St, Philadelphia, PA 19104 215-386-9601 Biweekly $ 58.00 per year.
News, politics of science, and career information for science professionals.

SCITECH BOOK NEWS
Book News, Inc., 5600 NE Hassalo St, Portland, OR 97213-3699 503-281-9230 Monthly $ 45.00 per year.
Reviews of new science books written primarily for librarians.

SKY & TELESCOPE
Sky Publishing Corp., 49 Bay State Rd, Cambridge, MA 02138 617-864-7360
Monthly $ 27.00 per year.
For professional and amateur astronomers. Articles and information on astronomy, space science, and telescopes.

SMITHSONIAN
Smithsonian Institution, 900 Jefferson Dr, SW, Washington, DC 20560-0001
202-357-2888 Monthly $ 22.00 per year.
History of sciences in all fields including social sciences.

TELESCOPE MAKING
Kalmbach Publishing Co., 21027 Crossroads Circle, Waukesha, WI 53187
800-446-5489
Covers optics, optical designs, machining, performance, adjustments with illustrations of complete telescope construction.

TODAY'S CHEMIST
American Chemical Society, Dept. 0011, Columbus, OH 43268-0011
203-226-7131 Bimonthly $ 17.00 per year.
Articles of news about current developments in chemistry.

TODAY'S SCIENCE ON FILE
Facts On File, Inc., 460 Park Avenue South, New York, NY 10016
800-322-8755 212-683-2244 Monthly
This news digest is published monthly and assembles *The Science News Digest* with cumulative index and glossary in a three ring binder for ready reference.

YEARBOOK OF SCIENCE AND THE FUTURE
Encyclopaedia Britannica, 310 S Michigan Ave, Chicago, IL 60604-4293
312-347-7000 Published annually in book form.
Annual developments in science summarized for the biological, physical and social sciences.

ZOO REVIEW
The Lincoln Park Zoological Society, 2200 N Cannon Drive, Chicago, IL 60614
312-935-6700 Free to members.
This magazine is published and distributed to members of the Society.

Chapter 11

Safety

Science Education Safety

Use this list to obtain various reference books and sources of materials and programs related to safety:

BEST'S SAFETY DIRECTORY
**A. M. Best Co., Inc., Ambest Rd, Oldwick, NJ 08858-9999   908-439-2200
Annually $ 39.00 per year.**
A buying guide and manual of safety practices safety supervisors.

CHEMICAL SAFETY
**Goldstein & Associates, 1150 Yale St, # 12, Santa Monica, CA 90403-4734
213-828-1309 Bimonthly**
Information on training, research and products.

142 Science Fun in Chicagoland

DRIVER'S POCKET GUIDE TO HAZARDOUS MATERIALS - 6TH EDITION
J. J. Keller & Associates, Inc., 145 W Wisconsin Ave, P. O. Box 368, Neenah, WI 54957-0368   800-558-5011   1989   289 pages
This comprehensive pocket reference was written for the transportation industry. This reference is often available at retail stores in truck stop centers.

GREATER CHICAGO SAFETY COUNCIL
1 N LaSalle, Chicago, IL 60602 312-372-9756
Ask about how schools may become members. Rental safety films are available to members at discount rates. Ask about monthly safety programs.

Do not be fooled! These chemistry teachers, called the Weird Science Kids, know safety procedures in the chemistry classroom. You may have seen Lee Marek squirt David Letterman with a fire extinguisher and cover him with shredded styrofoam. Photo courtesy of the Weird Science Kids.

HAZARDOUS WASTE RESEARCH AND INFORMATION CENTER - LIBRARY
Hazardous Waste Research and Information Center, 1 E Hazelwood Dr, Champaign, IL 61820   217-244-8989   217-333-8957
This library contains 10,000 books and 150 periodicals and is open to the public with no circulation privileges. Circulation through interlibrary loan is available.

HAZARDOUS WASTE RESEARCH AND INFORMATION CENTER
1808 Woodfield Dr, Savoy, IL 61874   217-333-8940
Ask for *Clearinghouse Publications List* of available free publications.

NATIONAL SAFETY COUNCIL - LIBRARY
Contact Robert J. Marecek, Manager, National Safety Council, 1121 Spring Lake Dr, Itasca, IL 60143-3201   708-285-1121
This library is open to the public and contains a comprehensive collection of 140,000 documents on health and safety. Fees apply to researchers requiring extensive assistance. It is best to telephone ahead for an appointment.

NATIONAL SAFETY COUNCIL
1121 Spring Lk Dr, Itasca, IL   708-775-2500, 1815 Landmeier Rd, Elk Grove Village, IL   708-981-0250
This nonprofit, nongovernmental organization offers a library, safety videos, training materials, and safety training instruction.

OCCUPATIONAL SAFETY AND HEALTH ADMINISTRATION (OSHA) TRAINING INSTITUTE
1555 Times Dr, Des Plaines, IL 60018   708-297-4810
Ask for catalog of audiovisual safety video cassettes for schools. This training center includes a safety equipment laboratory and a 3,000-book reference library.

THE POWER HOUSE - EDUCATIONAL PROGRAMS
Commonwealth Edison, 100 Shiloh Blvd, Zion, IL 60099   708-746-7492
Teachers and schools can request a program that visits schools, *Safety and Electricity*, for grades K-3. Ask for brochure on tours and educational programs.

TEACHING CHEMISTRY TO STUDENTS WITH DISABILITIES - 3RD EDITION
American Chemical Society, Chicago Section, 7173 N Austin, Niles, IL 60714   708-647-8405   1993   46 pages   Free
Ask for this excellent information and resource booklet.

THE TOTAL SCIENCE SAFETY SYSTEM - ELEMENTARY & SECONDARY EDITIONS
JaKel, Inc., 585 Southfork Dr, Waukee, IA 50263   515-225-6317
This computer software is an information data base about safety & science teaching, legal issues, safety assessment, and safety equipment resources. Request a brochure giving pricing. Developed by Dr. Jack Gerlovich, Assistant Professor of Science Education, Drake University.

Mail Order Safety Equipment Suppliers

ALDRICH SAFETY PRODUCTS
Aldrich Chemical Company, P. O. Box 2060, Milwaukee, WI 53201-0355 800-558-9160
Ask for 200-page catalog of laboratory equipment and supplies.

FLINN CHEMICAL CATALOG REFERENCE MANUAL
Flinn Scientific Inc., P. O. Box 219, 131 Flinn St, Batavia, IL 60510 800-452-1261   688 pages
This catalog lists new products, chemicals, chemical solution preparation, Apparatus & Laboratory Equipment, Books, Computer Software, Satety Storage Cabinets & Fume Hoods, Safety Supplies & Equipment, Right to Know Laws, Mystery Substance Identification, Chemical Inventory & Storage, and Chemical Disposal Procedures.

LAB SAFETY SUPPLY INC.
P. O. Box 1368, Janesville, WI 53547-1368 800-356-0783
Ask for 900-page catalog of laboratory equipment and supplies dedicated to personal environmental safety for industry, hazardous waste, and school science laboratories.

Chapter 12 Science Fairs

Science Fair Project Ideas

Contact your local public library. It should have an excellent collection of books on science fair project ideas and books on how to develop a project experiment. Some libraries conduct workshops on how to succeed with your science fair project.

1001 IDEAS FOR SCIENCE PROJECTS
by Marian Brisk 1992 242 pages. ($ 12.00 Showboard 800-927-3746)
This book lists supplies needed and time required and has ideas for all levels. One chapter is devoted to writing clear reports.

A SCIENCE PROJECT SURVIVAL GUIDE FOR KIDS AND ADULTS
by Edward Short 1991 47 pages. ($ 9.95 Showboard 800-927-3746)
This book helps students complete their project in simple, orderly tasks. For both elementary and high school grades.

Steps to a Winning Project Experiment

Good science fair projects include...

1. Making a study of other projects.

2. Choosing a topic idea or problem to study and stating the problem.

3. Stating the purpose, or value, of the project's experiment.

4. Conducting a literature search of related work in books and talking to professional experts.

5. Doing preparatory work, selecting materials, organizing the experiment.

6. Collecting raw experiment data.

7. Analyzing data, creating tables, diagrams, graphs, photographs and drawings.

8. Examining sources of error. Thinking. Drawing the best conclusions.

9. Preparing a written report and a brief oral presentation.

10. Constructing a science fair exhibit.

11. Participating in the science fair and the science fair judging process.

HAROLD WASHINGTON LIBRARY CENTER
Contact Deborah Mason, Science Fair Library Collection Librarian, 4th Floor, Science and Technology, 400 S State St, Chicago, IL 60605 312-747-4447
The Harold Washington Library Center has a large collection of books on science fair projects.

Chapter 12 - Science Fairs 147

NUTS & BOLTS
- A MATTER OF FACT GUIDE TO SCIENCE FAIR PROJECTS
by Van Deman and McDonald The Science Man Press, 1980. Out of print.
This well know book provides step-by-step instructions on how to develop a science fair project.

SCIENCE FAIR PROJECT INDEX 1973-1980
Akron-Summit County Public Library, Scarecrow Press, Inc., 1983. 729 pages Out of print.
This reference lists hundreds of science fair project ideas found in books and magazines published from 1973 to 1980. Check with your local public library to see this major reference.

SCIENCE FAIR PROJECT INDEX 1981-1984
by Bishop and Crowe Akron-Summit County Public Library, Scarecrow Press, Inc., 1986. 692 pages Out of print.
This reference lists hundreds of science fair project ideas found in 135 books and five magazines published from 1981 to 1984. Check with your local public library to see this major reference.

SHOW ME HOW TO WRITE
AN EXPERIMENTAL SCIENCE FAIR PAPER
by Judy Fisher Shubkagel 1993 26 pages ($ 10.95 Showboard 800-927-3746)
This reproducible workbook is recommended for Junior and Senior High.

SHOWBOARD - SCIENCE FAIR RESOURCE CATALOG
3702 W Sample St, South Bend, IN 46619 Midwest 800-927-3746, Nationwide 800-323-9189
Ask for 15-page catalog of materials for science fair projects and for running a local science fair, including project display boards, awards & ribbons, science fair quarterly newsletter, certificates of participation, idea books, and science fair videos.

SO YOU WANT TO DO A SCIENCE PROJECT!
by Joel Beller Prentice Hall Press, New York, 1982. 154 pages
This book covers all aspects of how to do a winning science fair project.

THE COMPLETE SCIENCE FAIR HANDBOOK
by Anthony Fredricks and Isaac Asimov 1990 ($ 9.95 Showboard 800-927-3746)
Recommended for grages 4-8, this book is filled with reproducible sheets with instructions on how to do an experiment and exhibit it at a science fair.

Teacher's Guide to Local Science Fairs

CHICAGO NON-PUBLIC SCHOOLS ANNUAL SCIENCE EXPOSITION
Sponsored by The Illinois Junior Academy of Science, Region 2, at the Museum of Science & Industry, Chicago, IL.
This regional science fair is held each year. For more information contact Ms. Julia Ferrari, Chairperson, Illinois Junior Academy of Science, Region 2, 708-253-7127.

CHICAGO PUBLIC SCHOOLS ANNUAL STUDENT SCIENCE FAIR
Sponsored by Chicago Public Schools Student Science Fair, Inc., P. O. Box 29546, Chicago, IL 60629 and the Department of Instructional Support, Chicago Public Schools, 1819 W Pershing Rd, Chicago, IL 60609 312-535-8850
Held annually in April, this exposition is sponsored by Argonne National Laboratory with support from The University of Chicago Board of Governors. Publications include the *Student Science Fair Program Guide to Projects, Information for Judges, Organizational Handbook for Administrators and Coordinators, and the Participant Handbook.*

ILLINOIS JUNIOR ACADEMY OF SCIENCE
Contact Ms. Judy Bonkalski, 1994-95 President, 708-739-1041, or Ms. Julia Ferrari, 1993-94 President, 708-255-8041. Founded in 1927.
Students from all over the state of Illinois present projects at regional expositions. The best of these projects are invited to the University of Illinois, Champaign/Urbana, in May of each year. Contact the Illinois Junior Academy of Science about how your students may participate.

INTERNATIONAL SCIENCE AND ENGINEERING FAIR
Science Service, Inc., 1719 N St, NW, Washington, DC 20036 202-785-2255
For over forty years Science Service has supervised the International Science and Engineering Fair. More than 800 students participate from over 20 different countries. Students in grades 9-12 are eligible and two student finalists are selected from each of the 415 regional science fairs.

MUSEUM OF SCIENCE & INDUSTRY - ANNUAL SCIENCE FAIRS
Contact Ed McDonald, Museum of Science & Industry, 57th St & Lake Shore Drive, Chicago, IL 60637 312-684-1414, ext 2423
Science Fairs are annually held at the Museum that encourage students to question, conduct research, and develop their own conclusions.

Chapter 12 - Science Fairs 149

This student knows that the science fair judge's interview can be a helpful exchange of ideas.

SCIENCE FAIRS AND PROJECTS - GRADES 7-12
Published and distributed by the National Science Teachers Association
800-722-NSTA 1988 72 pages $ 9.50
This book is for the teacher and explains all aspects of how to develop a successful science fair.

SCIENCE FAIRS AND PROJECTS - GRADES K-8
Published and distributed by the National Science Teachers Association
800-722-NSTA 1988 72 pages $ 9.50
This book is for the teacher and explains all aspects of how to develop a successful science fair.

Chapter 13

Toys

The word toy means fun to a child. Toys, and the boxes they come in, are all delightful interactive experiments for the child. Learning about the world can be thrilling and toys are one way a parent or a teacher can guide learning.

Good science toys are those which not only introduce science topics to the child but also allow the child to learn positive attitudes, to develop motor skills, to be creative, and to learn to think.

Science Toy Reference Books

EARTH-FRIENDLY TOYS
by George Pfiffner   John Wiley & Sons, Inc., New York 1994 128 pages $ 12.95
This book describes how to make toys and games from reusable objects. One learns recycling while making fun mechanical action science toys.

SCIENCE FARE
by Wendy Saul Harper & Row, Publishers, New York, 1986 Paperback, 295 pages
A classic illustrated guide and catalog of toys, books and activities about science for kids. The first nine chapters discuss science education for parents and teachers and the last eleven chapters describe specific resources by subject.

TEACHING SCIENCE WITH TOYS (TWO VOLUMES)
by the Center for Chemical Education, Miami University Middletown, 4200 E University Blvd, Middletown, OH 45042 $ 50.00 (Elementary School to Middle School)
This collection of over 80 chemistry and physics activities for grades K-9 was tested by classroom teachers from around the country.

TOYS IN SPACE: EXPLORING SCIENCE WITH THE ASTRONAUTS
by Dr. Carolyn Sumners, Project Director for the Toys in Space Program, NASA TAB Books, Division of McGraw-Hill, Inc., Blue Ridge Summit, PA 17294-0840 800-822-8158 1994 141 pages Hard cover $ 17.95, paperback $ 10.95 (800-722-NSTA $ 10.95)
Many mechanical action toys were taken on a NASA shuttle mission to observe their motion in weightless space. These toys, how they move on Earth, and what happened to them in space are described.

Educational Toy Loan Centers

Toys can be expensive and strain limited budgets. One option to this dilemma is the Lekotek toy library. Parents can expose their children to a large number of quality educational toys. Several Lekotek centers are located in the Chicagoland area.

Centers generally provide a monthly one hour session for the child after which the child takes home five toys. Often introductory tours for parents are scheduled each month. Most centers are open to everyone.

CLEARBROOK CENTER - LEKOTEK
Contact Sheila Lullo, Director, Clearbrook Center for the Handicapped, 3705 Pheasant Dr, Rolling Meadows, IL 60008 708-392-2812 Fee $ 65 per year.
Ask for descriptive brochure. This toy loan center for children with special needs is open to everyone in the Chicagoland area.

152 Science Fun in Chicagoland

NATIONAL LEKOTEK CENTER
Contact Rebecca Krook, Administrator, National Lekotek Center, 2100 Ridge Ave, Evanston, IL 60201   708-328-0001   Fee $180 per year.
Ask for descriptive brochure of the National Lekotek Center, a not-for-profit charitable organization. Also ask for a current list of over 20 Illinois Lekotek sites including Chicago, Chicago Heights, Flossmoor, Franklin Park, Libertyville, Lombard, Rolling Meadows, South Holland, and Tinley Park. Some centers have no fee and focus on special community needs.

Local Toy Stores Selling Science Toys

There are many fine retail toy stores in Chicagoland. See your telephone yellow pages under Toys-Retail for stores near you. The following list includes stores that science teachers recommend for the science toys sold there. They are popular among science teachers who use these toys in the classroom.

AMERICAN SCIENCE & SURPLUS
5696 Northwest Highway, Chicago, IL 60646   312-763-0313   and 1/4 mile east of Kirk Road, on Route 38, Geneva, IL 60185   708-232-2882   Mail order warehouse: American Science & Surplus, 3605 Howard St, Skokie, IL 60076   708-982-0870
These retail stores are an extensive resource of inexpensive, surplus science equipment. They are a favorite of science students, science teachers and do-it-yourself inventors. From electric wires and motors to test tubes and telescopes American Science & Surplus seems to have it all. Visit one of these stores and let your creative energy go wild. Ask for a mail order catalog.

BAREBONES
K106, Woodfield Mall, Schaumburg, IL 60173   708-413-2663
Although this store concentrates on toys and models about the human body and its anatomy, you will find fun oddities from geology and optical illustions.

CHICAGO KITE CO.
6 S Brockway, Palatine, IL 60067   708-359-2556
This retail store specializes in kites. It also has boomerangs and other science toys.

CONSTRUCTIVE PLAYTHINGS
5314 W Lincoln Ave, Skokie, IL 60076   708-675-5900
This parent/teacher retail store is filled with educational fun for the preschool and elementary school age child.

CUT RATE TOYS
5409 W Devon Ave, Chicago, IL 60646   312-763-5740
For over 40 years this retail store has sold major name toys at discount prices as well as surplus and discontinued items of all price ranges. Science principles and concepts hide in many of these toys. A large metal "Slinky" costs $ 1.59.

DEVON TOYS
2424 W Devon Ave, Chicago, IL 60659   312-973-0194
Former location of Cut Rate Toys. This retail store sells major name toys at discount prices.

DOOLIN AMUSEMENT SUPPLY CO.
511 N Halsted, Chicago, IL 60622 312-243-9424   Party supplies
A retail source of balloons and novelties for science experiments.

ENCHANTED KINGDOM
North Pier, 435 E Illinois St, Suite 268, Chicago, IL 60611   312-321-5464 800-758-9028
This retail store carries science toys among its many quality toys.

GALT TOYS
900 N Michigan Ave, Chicago, IL 60611   312-440-9550
This store is a retail distributor of Galt Toys.

J.C.'S KITES
197 Peterson Rd., Libertyville, IL 60048 708-816-9990
This retail store specializes in kites. It also has boomerangs and other science toys.

KAY BEE TOYS
(See the yellow pages of your telephone directory to locate one of the twenty stores in the Chicagoland area.)
This toy retailer carries toys that use science principles as well as basic science toys.

THE KITE HARBOR
North Pier, 435 E Illinois St, Chicago, IL 60611   312-321-5483
This retail store specializes in kites.   It also has science toys like balloon helecopters, boomerangs, the swinging wonder and model planes.

KOHL LEARNING STORE
165 Green Bay Rd, Wilmette, IL 708-251-7168
This retail store within the Kohl Children's Museum is open to the public and sells many quality educational science toys for the preschool and elementary school child.

154 Science Fun in Chicagoland

LIGHT WAVE
North Pier, 435 E Illinois St, Chicago, IL 60611 312-321-1123
This retail store specializes in holographic art. Holograms are three dimensional photographs where laser light produces interference patterns on film. Here science and art mix to create unique visual surprises. This store is like a museum with no admission fee.

MUSEUM OF SCIENCE & INDUSTRY BOOKSTORE & SHOP
Museum of Science & Industry, 57th St and Lake Shore Drive, Chicago, IL 60637 312-684-1414, ext 2780
This retail store has its largest display at the main entrance within the Museum of Science & Industry. It exclusively sells science books, toys and novelties.

NATURAL WONDERS
Fox Valley Mall, 2330 Fox Valley Center, Aurora 60504 708-820-9004; Stratford Square, 609 Stratford Square, Bloomingdale 60108 708-893-9803; and Oak Brook Center, 504 Oak Brook Ct, Oak Brook 60521 708-954-6613
Nature and science gift stores. Ask for brochure, *Our Discount for Teachers*, that provides a savings of 15 % on purchases for the classroom.

STANTON HOBBY SHOP INC.
4718 N Milwaukee Ave (near Lawrence Ave), Chicago, IL 60630 312-283-6446
This very large hobby store has everything for the young scientist: Estes rockets, dinosaur models, chemistry lab kits, Smithsonian kits, and human anatomy models.

STOREHOUSE OF KNOWLEDGE
2822 N Sheffield, Chicago, IL 60657 312-929-3932
This retail store for school supplies has an extensive section of books and materials on science for the preschool and elementary school age child. It has many science gift ideas.

TOY STATION
270 Market Square, Lake Forest, IL 60045 708-234-0180
This store sells many different science kits, including Educational Insights Science Kits, microscopes, telescopes and astronomy materials.

TOYS R US
(See the yellow pages of your telephone directory.)
A major toy retailer that carries toys using science principles.

TROST HOBBY SHOP
3111 W 63rd St, Chicago, IL 60629 312-925-1000
This store carries basic hobby needs as well as science model kits, chemistry sets and kites.

Chapter 13 - Toys 155

UNCLE FUN
1338 W Belmont, Chicago, IL 60657   312-477-8223
A unique toy store with hard to find inexpensive toys. They have "hand boilers" illustrating the science behind the drinking bird and many other toys demonstrating science concepts and processes. The name of this store says it all.

Science Toy Sources

Quality science toys exist, but are sometimes hard to locate. This list of science toy sources includes both retail mail order sources and wholesale sources. Contact the wholesale sources to learn about retail distributors in your area. You can order directly from retail sources by mail or by telephone.

This list was created from and limited to catalog descriptions. Source descriptions are not comprehensive lists of toys available from each source. This list represents some of the many sources of science toys.

A G INDUSTRIES
15335 N E 95th St, Redmond, WA 98052   206-885-4599   Manufacturer   Retail and Wholesale
Ask for 12-page catalog. Educational kits for constructing paper airplane gliders, boats and origami. Whitewings series includes the History of Passenger Planes, Future of Flight, History of Jet Fighters, Racers, Space Shuttle, Science of Flight. Display cases are also offered for home and retail stores.

ACE-ACME
4100 Forest Park, St. Louis, MO 63108-2899   800-325-7888   Importer, Manufacturer and Distributor   Retail
Ask for 110-page catalog filled with a very large variety of inexpensive toys and novelties that demonstrate natural phenomena. An excellent source for science teachers.

ANDY VODA OPTICAL TOYS
P. O. Box 23, Putney, VT 05346   802-387-5457   Manufacturer   Retail
Ask for brochure. Phenakistascope with six magic wheels, Thaumatrope, Couples spinning pictures, Flipbooks, Greeting Flipbooks, Make-it-yourself Zoetrope.

ARCHIE MC PHEE
P. O. Box 30852, Seattle, WA 98103   206-782-2344
Ask for 30-page catalog. Fun, inexpensive toys and novelties. Insects, bats, turtles, fish, dinosaurs, eye balls, beanie with propeller, iguanas, wall walker octopus. Retail store located at 3510 Stone Way N, Seattle, WA.

BLOCKS AND MARBLES BRAND TOYS, INC.
P. G. Fettig Catalog Company, P. O. Box 27, Russellville, IN 46175 800-972-6228 317-435-2734 Manufacturer and Distributor Retail
Ask for brochure and price list. Sets of wooden blocks with holes and grooves assemble to make creative paths for rolling and falling marbles.

CAPRICORN TOYS, INC.
P. O. Box 148366, Chicago, IL 60614 312-929-5953 800-642-8237 Manufacturer Retail and Wholesale
Ask for brochure. Cube puzzles: Little Genius (beginner), Happy Cube (intermediate), Marble Cube (advanced); Creatics: foam creatures that can be assembled different ways; Caramba 4: a game of strategy.

CHILDCRAFT EDUCATION CORP.
20 Kilmer Road, P. O. Box 3081, Edison, NJ 08818-3081 800-631-5652 Distributor Retail
Ask for the 200-page catalog. Math toys, aquariums, plant growing kit, seashells, root garden, ant farm, microscope, optical toys, magnetic toys, weather materials, globes. Educational materials for preschool and elementary school.

CONSTRUCTIVE PLAYTHINGS
1227 E 119th St, Gradview, MO 64030-1117 800-448-4115
Ask for 184-page catalog filled with educational fun for the preschool and elementary school age child including six pages of hands-on science materials.

COPERNICUS
394 New York Ave, Huntington, NY 11743 516-424-2173 Distributor Wholesale
Ask for eight-page catalog. Member of Museum Store Association. Geodome, Volcano, Starglow, Rootbeer Kit, boomerangs, fly back plane, balloon car, Make a Clock Kit, Rattle Back, Weather Keychain, large Swinging Wonder, Echo Mike, Radiometer, Drinking Bird, magnetic wheel on wire frame, earth balls, Tornado Tube, magnifying glasses, kaleidoscope, astronaut ice cream, hand boiler, Magic Garden, abacus, Large Dome Making Kit, glow paint, glow bugs, spiral timer, auto compass, luminous star finder, small swinging wonder.

CRAFT HOUSE CORP.
P. O. Box 10031, Toledo, OH 43699-0031 (No phone orders) Manufacturer Wholesale and Retail
Ask for 30-page catalog. National Audubon Society hobby painting kits: Poster sets on animals, Water Fowl Woodlike Sculptures, and Wilderness and Young Wildlife Paint by Number Series. Nature pictures in hobby kits of hanging, stained-glass like ornaments.

Chapter 13 - Toys 157

Each year the Museum of Science & Industry lights up with science fair activity.

CREATIVE PUBLICATIONS
5040 W 111th St, Oak Lawn, IL 60453-5008 800-624-0822 Retail
Ask for the 115-page catalog filled with elementary school educational materials for math, geometry, science, science measurement, and educational curricula. Also science posters.

CREATIVITY FOR KIDS
1802 Central Ave, Cleveland, OH 44115 Distributor Wholesale
Ask for brochure. Hands On Science kits, Fun with Nature kit, Photography kit, Kitchen Chemistry kit.

CURIOSITY KITS
P. O. Box 811, Cockeysville, MD 21030 410-584-2605 Manufacturer Retail
Ask for 16-page catalog. Science curiosity kits include Wildlife Masks, The Bluebird House, The Insect House, Butterfly and Moth Collection, Skygazer's Mobile, Kaleidoscope, Seasonal Gathering Basket, Semiprecious Stone Jewelry.

158 Science Fun in Chicagoland

DAMERT COMPANY
2476 Verna Court, San Leandro, CA 94577   510-895-6500   800-321-3722
Manufacturer and Distributor   Retail and Wholesale
Ask for 20-page catalog. Science toys include 3-D Slide Puzzles, Jungle Bungles puzzles, Concentra puzzle, Tiazzle Puzzles, Master Triazzles, coffee mugs, bulletin boards, many science mobiles, StarShines astronomy stickers, diffraction toys, Laser Top, Spiral Mobiles, Turbo Sparkler YoYo, liquid crystal toys and novelties, Little Critter Kaleidoscope, Butterflies of the World, bird feeder kit, Zoetrope, Vector Flexor, Echo Rocket, Spacephones, Tornado Tube, science charts and posters.

DELTA EDUCATION, INC.
P. O. Box 3000, Nashua, NH 03061-3000 800-442-5444 Manufacturer Retail
Ask for 60-page *Hands-On Science Catalog*. This catalog is filled with science kits, toys and elementary school educaional materials. Also ask for information about *Delta Science Modules, SCIS3,* and *ESS*. These three hands-on programs are available through Delta Education.

DESIGN SCIENCE TOYS LTD.
1362 Route 9, Tivoli, NY 12583 800-227-2316 Manufacturer Retail
Ask for 20-page catalog. Tensegritoy construction toy, Stik-Trix construction puzzle, Roger's Connection magnetic sculpture, Hoberman Sphere, Celestial Orb, Fee Bee WeeBee flexable toy for newborns, Vexahedron, Dodeca puzzle, Rhoma puzzle, Rhomblocks construction blocks, Quad-Rhom with 4-D axis, Polygonzo, IcosaFlex, Cuboactaflex, The Star Series of transformable wire structures, Quix Series, Vectorsphere, Octabug, Synergy Ball.

EDMUND SCIENTIFIC COMPANY
101 E Gloucester Pike, Barrington, NJ 08007-1380 609-573-6250
Ask for 220-page catalog. Since 1942 this well known scientific optical supplier also sells many other items including lasers, microscopes, camera/monitor systems, science classroom anatomy models, nature kits, laboratory safety equipment, balances, weather instruments, timers, magnets, small motors & pumps, robot kits, earth science kits, telescopes, museum animal replicas, and unique classroom materials for teachers.

EDUCATIONAL DESIGN, INC.
345 Hudson St, New York, NY 10014-4502 212-255-7900   800-221-9372
Manufacturer Wholesale
Ask for 10-page catalog. Crystal Radio Videolab, Electric Motor Videolab, Electro-Magnetix Video Lab, Electricity Videolab, Minilabs on science subjects, Toplabs Big Labs each with several experiments, Powertech series of labs on technology, Superlabs each with several experiments. For Chicagoland retail distributor information contact: Eichas, Strickland Group at 708-740-8326.

EDUCATIONAL INSIGHTS
19560 S Rancho Way, Dominguez Hills, CA 90220   Manufacturer   Retail and Wholesale

Ask for 20-page catalog. GeoSafari computer games, Mini-Museums, Mystery Rock, Exploring Ecology, Natural Collections, Mysteries of Light, Mysteries of Magnetism, Adventures in Science series has 12 different projects, Science Safari Stickers, Fantastic Cards on science subjects, Animal Big Books 14" x 20", Bug Viewers, Dino Checkers, Dino Tic Tac Toe.

EDUCATIONAL TOYS, INC.
P. O. Box 630882, Ojus, FL 33163-0882   800-881-1800   Distributor   Retail

Ask for 30-page catalog. Wild Animal Families models, dinosaur and anatoy puzzles, dinosaur skeleton kits, Farm Life animal models, Monterey Bay Aquarium sea life models, space and mineral materials, The Carnegie Collection of dinosaur models, animal games, endangered animal puzzles, reptile models, Bug Jar, insect posters, toy microscopes and binoculars, Magnetic Marbles, Newton's Cradle swinging balls, kaleidoscope, models of the Rainforest Poison Dart Frog.

ESTES INDUSTRIES
1295 H Street, Penrose, CO 81240   719-372-6565

Ask for catalog on school letterhead. Supplies model rockets, engines and accessories.

EXPLORATORIUM STORE
3601 Lyon St, San Francisco, CA 94123   415-561-0393   800-359-9899

Ask for the 30-page *Exploratorium To Go!* Catalog that is filled with quality science toys and books, including the following toys: Megabubbles Kit, The Kaleidoscope Book and Kit, Zoetrope, Wild Wood, Magnetron, Gyros, Curiosity Box, Eagle Microscope, Mirage Maker, Micro-Bank, Erector Sets, Paradox 3-D Jigsaw Puzzle, and Ellipto.

EXPLORATOY
19560 S Rancho Way, Dominguez Hills, CA 90220   310-884-3490   800-995-9290   Manufacturer   Wholesale

Ten-page catalog. Beakman's World Inquizator computerized quiz machine, Early Start learning machine, Critter Carnival insect house, Creature Catcher, The Antworks, Bug Pals, Riddle Rocks, Explorascope microscope, Test Tube Science in six science packs, Cosmic Observing Station telescope, Star Tower toy planetarium.

GEOSPACE PRODUCTS COMPANY
1546 N W Woodbine Way, Seattle, WA 98177   800-800-5090   Manufacturer   Wholesale and Retail

Magnetic marble toys, magnetic levitation games, and magnetic building sets.

160 Science Fun in Chicagoland

HOMECRAFTERS MANUFACTURING
1859 Kenion Point, Snellville, GA 30278 404-978-3012 Manufacturer Wholesale and Retail
Ask for brochure describing "I Made My...Periscope!" kit and "I Made My...Kaleidoscope!" kit.

HORTICULTURAL SALES PRODUCTS
P. O. Box 251, Ft Lee, NJ 07024 201-585-7077 Manufacturer Wholesale
Manufacturer of Root-Vue-Farm. Watch carrots, radishes and onions take form underground through a glass window.

IDEAL SCHOOL SUPPLY COMPANY
11000 S Lavergne Ave, Oak Lawn, IL 60453 800-845-8149 Distributor Retail
Ask for the 50-page teacher catalog. Preschool and elementary school science measurement materials, chemistry experiment beakers and test tubes, equilateral prisms, physics pulleys, thermometers, classroom science kits, magnetic toys, natural science materials.

JAMES GALT COMPANY, INC.
63 N Plains Highway, Wallingford, CT 06492 203-265-7222 800-966-GALT Distributor Wholesale and Retail
Math toys, AquaPlay water toys, Carnegie Collection of dinosaur models, optical toys, dolphin and aquatic life models, Super Bug Jar, magnetism toys, color addition toys.

KADON ENTERPRISES, INC.
1227 Lorene Dr, Suite 16, Pasadena, MD 21122 410-437-2163 Manufacturer Retail
Ask for 15-page catalog of *Gamepuzzles: for the Joy of Thinking*. This company specializes in sophisticated games and puzzles for the creative thinker.

KIPP BROTHERS, INC.
240-242 S Meridian St, P. O. Box 157, Indianapolis, IN 46206 800-428-1153 Importers Retail and Wholesale
Ask for 224-page catalog. Established in 1880 this distributor specializes in inexpensive toys, novelties, carnival and party items for quantity, dozen purchases. Science items include dinosaur tattos, animal sounds, musical toys, tops, magnetic wheels, rubber and foam balls, kaleidoscopes, bird gliders, solar radiometer, telescopes, boomerangs, magnetized marbles, flying toys, kazoos, magnifying glasses, museum quality dinosaurs, and many, many more.

KLUTZ
2121 Staunton Court, Palo Alto, CA 94306   415-424-0739   Manufacturer and Distributor   Retail
Ask for the 70-page Flying Apparatus Catalogue. Really fun toys and novelties. Amazon Worms, Smartballs, Smartrings, The Explorabook - a kids science museum in a book, ExploraCenter, Backyard Weather Station kit, Mega-Magnet Set, Backyard Bird Book with bird caller, The Aerobie Orbiter, Rubber Stamp Bug Kit, Vinyl Vermin, Kids Gardening Guide, World Record Paper Airplane Kit, The Arrowcopter, Megaballoons, Bubble Book, Zoetrope, juggling materials.

KOLBE CONCEPTS, INC.
P. O. Box 15667, Phoenix, AZ 85060   602-840-9770   Manufacturer   Retail
Think-ercises, Glop Shop - inventor's assortment, Go Power - science experiments, Using Your Senses, Solar Power Winners - experiment book, Decide & Design - inventor's book.

LAKESHORE LEARNING MATERIALS
2695 E Dominguez St, Carson, CA 90749   800-428-4414   Manufacturer and Distributor   Retail
Ask for 200-page catalog. The catalog of this major distributor of elementary school learning materials has eight pages of science materials.

LEARNING RESOURCES, INC.
675 Heathrow Drive, Lincolnshire, IL 60069   708-793-4500   800-222-3909
Manufacturer   Wholesale
Educational materials. Mini-Dinos Activity Kit, Story Puzzles with Animals, math and science measurement materials, geometry shapes, wood base ten blocks, thermometers, microscopes, prehistoric animal models, color paddles, Power of Science kits, technology kits, nature kits, measurement sets.

LEGO DACTA
- THE EDUCATIONAL DIVISION OF LEGO SYSTEMS, INC.
555 Taylor Rd, P. O. Box 1600, Enfield, CN 06083-1600   800-527-8339
Manufacturer   Retail and Wholesale
Gear, lever and pulley toys, Technic classroom kits, Technic control centers, teacher's guide books, Pneumatics, Logowriter Robotics for Apple and MS-DOS, Control Lab for Apple and MS-DOS.

LIBBY LEE TOYS, INC.
7650 School Road, Cincinnati, OH 45249-1592   513-489-8080   800-542-2953
Manufacturer   Wholesale
Ask for 20-page catalog. Plantsters Greenhouses in various elementary school educational kits are for growing plants indoors with step by step instructions for the child.

162 Science Fun in Chicagoland

NATURE'S TOYLAND
Subsidiary of Penn-Plax, Inc., 720 Stewart Ave, Garden City, MY 11530
516-222-1020   Manufacturer   Wholesale
This manufacturer of pet care products makes educational kits, including Tweety-Your First Bird Cage Kit, Tom and Jerry Hamster/Gerbil Home, The Little Mermaid Goldfish Aquarium and Collection Tank, The Little Mermaid Hermit Crab World, Ninja Turtles Collection Play Tank, Tweety-My First Bird Watching Kit, Bugs Bunny Rabbit/Guinea Pig Cage and Small Animal Habitats for hamsters and gerbils. See your local pet care retail store.

Pets are not toys, but Nature's Toyland's educational pet care kits help children learn about pet care and about nature. **Photo courtesy of Nature's Toyland.**

NATUREPRINT PAPER PRODUCTS
P. O. Box 314, Moraga, CA 94556 510-284-3115 Manufacturer Retail
Natureprint paper and transparencies. This sun-sensitive paper exposes in direct sunlight to create white on blue prints of leaf outlines or animal picture transparencies. Expose for 2-3 minutes and then develop in tapwater in seconds.

ORIENTAL TRADING CO., INC.
P. O. Box 3407, Omaha, NE 68103-0407 Orders 800-228-2269, Customer Service 800-228-0475 Distributor Wholesale and Retail
Ask for catalog. Source of inexpensive novelties, including magnifying glasses, plastic tops, mini reflectors, balloons, balloon helicopters.

OWI INCORPORATED
1160 Mahalo Place, Compton, CA 90220-5443 310-638-4732 Manufacturer Wholesale
Ask for brochure on Robotics and for information on retail distributors. This manufacturer makes several different robotic kits requiring different levels of assembly sophistication. For age nine years to adult.

PEELEMAN/MC LAUGHLIN ENTERPRISES, INC.
4154 S 3rd West, Murray, UT 84107 800-779-2205 Manufacturer Retail
Ask for 16-page catalog. Science toys include Children's Solar System, Rain Forest in a Case, and Ocean in a Case.

PHYSICS OF TOYS DEMONSTRATION SET
71938-02 $ 159.00 and *Physics Fun and Demonstrations Manual* # 58225 $ 11.95 CENCO, 3300 Cenco Parkway, Franklin Park, IL 60131-1364 800-262-3626
This set of 23 familiar toys demonstrate physical principles as explained in the accompanying manual.

PLAY VISIONS
1137 N 96th St, Seattle, WA 98103 800-678-8697 206-524-2774 Distributor Wholesale
Ask for 50-page catalog. Telescopes, optical toys, Giant Rainforest Insects, reptiles & amphibians, Habitat nature model sets, dinosaur toy sets, vinyl Earth balls, Earth Squish Balls, Native American Arrowheads, and many inexpensive novelty items.

PLAY-TECH, INC.
139 Harristown Rd, Glen Rock, NJ 07452 201-670-1655 Manufacturer and Distributor Wholesale
Ask for brochures. Mr. Wizard's Science Secrets sets, Motorized Capsela Science Discovery Series construction sets, X-Men Magnifiers binoculars, Thomas The Tank Engine Teaching Toys series.

164 Science Fun in Chicagoland

SCHYLLING ASSOCIATES, INC.
44 Mitchell Road, P. O. Box 667, Ipswich, MA 01938   508-356-1600   Importer and Distributor   Retail
Ask for 16-page catalog. Wind-up walking dinosaurs, space tops, glow in the dark 26" Earth poster, volanco kit, cylindrical mirror fun scopes, star gazer glow in the dark solar system, toy periscope, solar energy kit, orbit ring toy with ball inside hoop, Lehman tin/mechanical robot and space toys.

SELSI COMPANY, INC.
40 Veterans Blvd, P. O. Box 497, Carlstadt, NJ 07072   201-935-0388   800-275-7357   Manufacturer   Wholesale
Quality binoculars, telescopes, student microscope sets, magnifiers, toy kaleidoscopes, glass prisms, student magnets, compasses, barometers, altimeters, metal detectors.

SMALL WORLD TOYS
5711 Buckingham Parkway, Culver City, CA 90230   800-421-4153   Distributor   Wholesale
Gyroscopes, Gravity Graph, inflatable globes, Backyard Scientist, mineral sets, Bug World, dinosaur skeleton kits, Polyopticon optical toy kits, Bug Hotel, magnetic toys, Gigantic Glow Stars, dinosaur models, origami kits, Whirlybirds, Newton's Yo-Yo, Finger Tops, Astronaut Food, animal sets, magnifier toys, Sparkling Wheels, Relaxable Globe Balls.

SOMMERVILLE HOUSE
3080 Yonge St, Suite 5000, Toronto, Canada M4N 3N1   416-488-5938   Manufacturer   Wholesale
The Bones Book with plastic skeleton, dinosaur books with plastic skeletons, The Environmental Detective Kit. Books packaged with toy models.

SUMMIT LEARNING
P. O. Box 493, Ft. Collins, CO 80522   800-777-8817   Retail
Ask for the 75-page catalog filled with educational materials for math and science including the following categories: Linear Tools; Volume and Capacity; Weights and Measures; Time and Temperature; Problem-Solving; Estimation; Graphs; Probability; Earth Science; Astronomy; and Science and Nature.

TAMIYA AMERICA, INC.
2 Orion, Aliso Viejo, CA 92656-4220   Manufacturer   Wholesale
Ninety-page catalog of plastic scale model kits includes Dinosaur Diorama Series, Educational Construction series of walking dinosaurs and action robots, Construction Materials Technicraft Series of electrical, solar and mechanical components. See your local hobby store.

TYCO TOYS, INC.
6000 Midlantic Drive, Mount Laurel, NJ 08054 609-840-1327 **Manufacturer Wholesale**
Chemcraft chemistry sets, Doctor Dreadful food labs, Smithsonian Institution dinosaur sets.

UNCLE MILTON INDUSTRIES, INC.
3555 Hayden Ave, Culver City, CA 90232-0246 310-559-1566 **Manufacturer Wholesale**
Seven-page catalog contains ant farms (Milton Levine invented the ant farm in 1956), Pocket Museums, Fossil Hunt, Krazy Klowns, Light-Up Critter City, BugJug, Star Theater home planetarium, Super GeoScope microscope, Hydro Greenhouse, Rock & Mineral Hunt.

UNIVERSAL SPECIALTIES CO., INC.
7355 W Vickery Blvd, Fort Worth, TX 76116 817-738-5299 (Order at Alex-Panline USA, Inc., Englewood, NJ 800-666-2539) **Distributor Wholesale**
Ask for 30-page catalog. This distributor of inexpensive novelty items sells many toys about science or that demonstrate natural phenomena.

Chapter 14

Video

Film and Video Sources

BLOCKBUSTER VIDEO
(See telephone yellow pages for location near you.)
Video rental. See Science and Nature Section which includes Audubon videos, Cousteau videos, National Geographic Video series, NOVA series, and PBS Home Video.

COLLEGE VIDEO CORPORATION
1106 Laxton Rd, Suite B, Lynchburg, VA 24502   800-852-5277
Students enrolled in college credit video courses, (such as telecourses offered at the Center for Open Learning, City Colleges of Chicago, 30 E Lake St, 11th Floor, Chicago, IL 60601, via WYCC/Chicago Channel 20 television, 312-553-5970), may rent the entire educational video series from College Video Corporation for a period of four months. Inquire about available video series and rental rates.

EDUCATIONAL ACTIVITIES, INC.
P. O. Box 392, Freeport, NY 11520   800-645-3739
Ask for 20-page catalog of video cassettes and videodiscs, including science videos.

EDUCATORS GUIDE TO FREE FILMS
Educators Progress Service, Inc., 214 Center St, Randolph, WI 53956-1497   414-326-3126   1994 $ 28.95
This book lists and describes free films available to educators.

EDUCATORS GUIDE TO FREE SCIENCE MATERIALS - 35TH EDITION
edited by Mary H. Saterstrom   Educators Progress Service, Inc., 214 Center St, Randolph, WI 53956-1497   414-326-3126   1994   296 pages $ 27.95
This book lists and describes free science materials, including films, filmstrips, slides, audiotapes, and printed materials by category of science subject area.

EDUCATORS GUIDE TO FREE VIDEOTAPES
Educators Progress Service, Inc., 214 Center St, Randolph, WI 53956-1497   414-326-3126   1994 $ 26.95
This book lists and describes free videotapes available to educators.

HAWKHILL ASSOCIATES, INC.
125 E Gilman St, P. O. Box 1029, Madison, WI 53701-1029   800-422-4295
Ask for 15-page catalog of video tapes on science topics.

INSIGHT MEDIA
2162 Broadway, New York, NY 10024   212-721-6316
Ask for brochure of videocassettes on all science topics for the high school classroom.

INVENTING THE FUTURE: AFRICAN-AMERICAN CONTRIBUTIONS TO SCIENTIFIC DISCOVERY AND INVENTION
American Chemical Society, 1155 16th St, NW, Washington, DC 20036   202-452-2113   VHS Video, 14 minutes. $ 10.00
Overview of historic African science including almanacs, astronomical measurements, and railroad communication covering the mid 1970's to the mid 1900's. This video includes a 12-page teacher's guide of biographical material as well as hands-on science activities.

NSTA SCIENCE EDUCATION SUPPLIERS
A Supplement to: Science & Children, Science Scope, and The Science Teacher, National Science Teachers Association, 1840 Wilson Blvd, Arlington, VA 22201-3000 800-722-NSTA Published annually. 122 pages $ 5.00 per copy.
List of educational video producers and distributors. The most current and comprehensive list of manufacturers, publishers and distributors of science education materials. See Media Producers. In 1994, 257 media producers were listed.

PHYSICS CURRICULUM & INSTRUCTION
22585 Woodhill Dr, Lakeville, MN 55044 612-461-3470
Ask for 10-page catalog of physics demonstrations and concepts on videocassette and laserdisc.

SCIENCE BOOKS & FILMS
American Association for the Advancement of Science, 1333 H St, NW, Washington, DC 20005 Monthly
This periodical reviews scientific accuracy and presentation of print, audiovisual, and electronic resources intended for educational use.

TRACING THE PATH: AFRICAN-AMERICAN CONTRIBUTIONS TO CHEMISTRY IN THE LIFE SCIENCES
American Chemical Society, 1155 16th St, NW, Washington, DC 20036 202-452-2113 VHS Video, 18 minutes. $ 10.00
Overview of historic African science and technolog with an emphasis on traditional African healing. This video includes a 12-page teacher's guide of hands-on activities, discussion topics, and demonstrations.

UNIVERSITY OF CALIFORNIA EXTENSION CENTER FOR MEDIA AND INDEPENDENT LEARNING
2000 Center St, Fourth Floor, Berkeley, CA 94704 510-642-0460
Request *Film & Video Sales Catalog* and *Film/Video Rental Catalog* offering videos like *When the Bay Area Quakes*, the series *General Chemistry Laboratory Techniques*, and the series *Understanding Space and Time*.

UNIVERSITY OF ILLINOIS FILM/VIDEO CENTER
1325 S Oak St, Champaign, IL 61820 800-367-3456
Call and ask about videos available in science, for example ask for the *Catalog of Solid Waste Mangement Video Resources*.

Chapter 14 - Video

MR. WIZARD INSTITUTE
A Division of Prism Productions, 44800 Helm St, Plymouth, MI 48170 313-459-6668
Ask for brochure that describes *Mr. Wizard's World Science Video Library*, available on videocassettes, Mr. Wizard's Science Secret Kits, Teacher to Teacher...with Mr. Wizard, and Science Twenty Video Tapes. The very best of Mr. Wizard's World, the popular children's television program seen on *Nichelodeon*.

ZTEK CO.
P. O. Box 1055, Louisville, KY 40201-1055 502-584-8505
Ask for 30-page catalog of educational multimedia on laserdiscs.

Science Programs on Television

Check local television programming for these educational science programs. Chicago has two Public Television System (PBS) stations: WTTW/Chicago Channel 11 and WYCC/Chicago Channel 20.

3-2-1 CLASSROOM CONTACT (SCIENCE, GRADES 4-6)
PBS Elementary/ Secondary Learning Services, 1320 Braddock Place, Alexandria, VA 22314-1698 800-228-4630 **for single copies of the guide for $ 15.35.**
This PBS series about the curiosity and mystery of science topics is available in a special classroom edition that comes with an easy-to-use teacher's guide.

THE ADVENTURES OF HYPERMAN
This television program planned for ages 8-11, focuses on a superhero who promotes getting interactive with CD-ROM computers. Check local television station programming.

BILL NYE THE SCIENCE GUY
KCTS/Seattle, 401 Mercer St, Seattle, WA 98109 206-624-1915
The wild and wacky style of this program provides an entertaining approach to science for the younger viewer. Ask for 22-page teacher's guide and educational materials available to schools. Inquire about videotapes containing the best episodes: Disney Educational Productions, 500 S Buena Vista St, Burbank, CA 91521-6677 818-567-5684

THE CHALLENGE OF THE UNKNOWN
(MATHEMATICS, GRADES 4-12)
PBS Elementary/ Secondary Learning Services, 1320 Braddock Place, Alexandria, VA 22314-1698
This PBS series is about applied mathematics and problem solving designed for use in elementary, junior and senior math and science classrooms. A 288-page teacher's guide can be purchased for $ 6.00 from Karol Media, 350 N Pennsylvania Ave, Wilkes Barre, PA 18773 717-822-8899

FUTURES-1 WITH JAIME ESCALANTE
(MATH & SCIENCE, GRADES 7-12)
PBS Elementary/ Secondary Learning Services, 1320 Braddock Place, Alexandria, VA 22314-1698 703-739-5038 for $ 3.00 Teacher's Guide.
This PBS series shows how math is applied in the real worlds of fashion, space exploration, sports, and other careers. A 42-page teacher's guide is available for $ 3.00 (minimum order of 10).

FUTURES-2 WITH JAIME EXCALANTE
(MATH & SCIENCE, GRADES 7-12)
PBS Elementary/ Secondary Learning Services, 1320 Braddock Place, Alexandria, VA 22314-1698
This PBS series is a continuation of Futures-1.

ICEWALK (ENVIRONMENTAL SCIENCE, GRADES 7-12)
PBS Elementary/ Secondary Learning Services, 1320 Braddock Place, Alexandria, VA 22314-1698 703-739-5038 $ 3.00 Teacher's Guide.
This PBS series is designed to inspire young people to work to save the environment. A teacher's guide is available for $ 3.00 (minimum quantity of 30).

NATIONAL AUDUBON SOCIETY SPECIALS
PBS Elementary/ Secondary Learning Services, 1320 Braddock Place, Alexandria, VA 22314-1698
This PBS series for grades 4-12 examines the wonder and beauty of nature's rarest creatures. Each program on VHS costs $ 49.95, telephone 800-344-3337. Each program is accompanied by a a 1-5 page teacher resource.

THE NEW EXPLORERS
produced by Bill Kurtis
See WTTW/Chicago Channel 11 for current programs. For information on educational materials and videos telephone 800-621-0660. This series investigates current science discoveries featuring the scientists who make these discoveries. Ask for *The Science Explorers Program Information Packet*, including a list of regional facilities, by writing to LaVonia Ousley, Division of Educational Programs, Argonne National Laboratory, 9700 S Cass Ave, Bldg 223, Argonne, IL 60439.

NOVA
NOVA Guide, Educational Print and Outreach, WGBH, 125 Western Ave, Boston, MA 02134
Over the past 20 years the NOVA television series has covered current topics in the sciences. A 30-page teacher's guide is available upon request.

PYRAMID (INTERDISCIPLINARY, GRADES 5-9)
PBS Elementary/ Secondary Learning Services, 1320 Braddock Place, Alexandria, VA 22314-1698
This PBS program discovers the mysteries about the construction and times of the ancient Egyptian pyramids.

RACE TO SAVE THE PLANET (SCIENCE, GRADES 7-12)
PBS Elementary/ Secondary Learning Services, 1320 Braddock Place, Alexandria, VA 22314-1698
This PBS series examines major environmental questions using a case study approach.

READING, THINKING, & CONCEPT DEVELOPMENT (PROFESSIONAL DEVELOPMENT, TEACHERS GRADES K-12)
PBS Elementary/ Secondary Learning Services, 1320 Braddock Place, Alexandria, VA 22314-1698
This PBS series for teachers and administrators is designed to help students with higher order thinking skills.

SCIENCE POWER
Chicago Cable Television, Channel 21. by Dr. Diane Schiller, Loyola University Chicago and Philip C. Parfitt, The Chicago Academy of Sciences. On the air September-May, Wednesdays, 7:30 p.m. to 8:30 p.m.
This interactive program allows students, elementary through middle school, to phone and participate during live broadcasts about science. They answer questions and ask questions about the science on the air. Here everyone has fun as they learn about science.

SCIENTIFIC AMERICAN FRONTIERS
Scientific American Frontiers School Program, Connecticut Public Television, P. O. Box 260240, Hartford, CT 06126-0240 800-523-5948
This television series takes you on excursions into the furthest realms of scientific investigation. A 15-page teacher's guide is available upon request.

SECOND VOYAGE OF THE MIMI (SCIENCE, GRADES 4-8)
PBS Elementary/ Secondary Learning Services, 1320 Braddock Place, Alexandria, VA 22314-1698 703-739-5038
This PBS series combines dramatic episodes and documentary expeditions to develop mathematical and scientific concepts. A teacher's guide is available for purchase.

THE SECRET OF LIFE
Educational Print and Outreach, WGBH, 125 Western Ave, Boston, MA 02134; To purchace videos: The WGBH Collection, P. O. Box 2053, Princeton, NJ 08543 800-828-WGBH $ 89.95 per program.
This television series focuses on how DNA research is helping scientists to develop diagnostic tools and treatments for inherited diseases. A 33-page teacher's guide is available upon request.

TEAMS
Los Angeles County Office of Education, 9300 Imperial Highway, Downey, CA 90242-2890 213-922-6603
Telecommunication Education for Advances in Mathematics and Science (TEAMS) is a Federally funded national school consortium. They have a network of mathematics and science programs that are broadcast from consortium schools. Ask for an information packet.

UNDER THE MICROSCOPE
(SCIENCE TEACHER STAFF DEVELOPMENT, GRADES 3-7)
PBS Elementary/ Secondary Learning Services, 1320 Braddock Place, Alexandria, VA 22314-1698 703-739-5038 for $ 2.00 Teacher's Guide.
This PBS program describes what teachers want to know when teaching elementary school science. A teacher's guide is available for $ 2.00 (minimum order of 10).

VOYAGE OF THE MIMI, THE (SCIENCE, GRADES 4-8)
PBS Elementary/ Secondary Learning Services, 1320 Braddock Place, Alexandria, VA 22314-1698 703-739-5038
This PBS series combines dramatic episodes and documentary expeditions to develop mathematical and scientific concepts. A teacher's guide is available for purchase.

INDEX

Index

1001 IDEAS FOR SCIENCE PROJECTS 145
175 AMAZING NATURE EXPERIMENTS 8
2061
    PROJECT 2061 16
    SCIENCE FOR ALL AMERICANS 16
3-2-1 CLASSROOM CONTACT 169
3-2-1 CONTACT 123
57th STREET BOOKS 19
700 SCI EXPERIMENTS FOR EVERYONE 8
A CLASS ACT 127
A G INDUSTRIES 155
A T & T BELL LABORATORIES
    - INDIAN HILL LIBRARY 93
A T & T INFORMATION RESOURCE CTR 93
AAAS PRESS 23
AAPT PRODUCTS CATALOG 37
ABBOTT LABORATORIES
    - INFORMATION SERVICES 93
ABRAMS PLANETARIUM CALENDAR 123
ACADEMIC INC. 39
ACCELERATOR NEWSLETTER 126
ACCESS 2000 44, 78
ACE-ACME 110, 155
activities 8
    1001 IDEAS FOR SCI PROJ 145
    175 AMAZING EXPERIMENTS 8
    700 SCIENCE EXPERIMENTS 8
    ASTRONOMY FOR EVERY KID 8
    CLASSROOM CREATURE 8
    EARLY CHILDHOOD 9
    EARTH SCIENCE FOR 9
    EVERYDAY SCIENCE 9
    EXPERIMENTAL SCIENCE 9
    EXPLORATORIUM SNACKBOOK 10
    HELPING YOUR CHILD 10
    ICE PICKS 3
    INVITATIONS TO SCIENCE 10
    KITCHEN SCIENCE 11
    NSTA MEMB 26
    PHYSICS EXPERIMENTS 11
    SCIENCE ACTIVITIES 130
    SCIENCE AND CHILDREN 130
    SCIENCE IS... 12
    SCIENCE ON A SHOESTRING 12

SCIENCE STARTERS 13
SCIENCEWORKS 13
STRING AND STICKY TAPE EXP 13
SUPERMARKET SCIENCE 11
THOMAS EDISON BOOK 14
ACTIVITIES INTEGRATING MATHEMATICS
    AND SCIENCE 110
ADAM SOFTWARE 32
ADLER PLANETARIUM 65
    LIBRARY 93
    DEPARTMENT OF EDUCATION 44
    PLANETARIUM SHOP 19, 104
ADVENTURES OF HYPERMAN 169
African science
    INVENTING THE FUTURE 167
    MULTICULTURALISM 11
    TRACING THE PATH 168
AIMS 110
AIR & SPACE /SMITHSONIAN 133
AIR POLUTION BBS 32
ALDRICH SAFETY PRODUCTS 144
ALLIANCE FOR ENVIR EDUCATION 84
ALTIMETER 88
amateur radio
    AM RADIO RELAY LEAGUE 44
AMERICAN ASSOC FOR THE ADVANCEMENT
    OF SCIENCE 84
AMERICAN ASSOC OF PHYSICS TEACHERS
    23, 37, 84
AMERICAN BIOLOGY TEACHER 127
AMERICAN CHEMICAL SOCIETY 84
    CHICAGO SECTION 78
    TEACH RESOURCES CATALOG 110
AMERICAN GEOLOGICAL INSTITUTE 24, 84
AMERICAN JOURNAL OF PHYSICS 127
AMERICAN KITE 123
AMERICAN MEDICAL ASSOCIATION 94
AMERICAN NUCLEAR SOCIETY
    LIBRARY 94
AMERICAN RADIO RELAY LEAGUE 44, 84
AMERICAN SCIENCE & SURPLUS 104, 152
AMERICAN SCIENTIST 133
AMOCO CORPORATION - CENTRAL
    RESEARCH LIBRARY 94

176 Science Fun in Chicagoland

AMOCO CORP - LIBRARY/INFO CENTER 94
AMUDENSEN HIGH SCHOOL 42
AMUSEMENT PARK PHYSICS 8, 63
anatomy
    ADAM SOFTWARE 32
    BAREBONES 106
    DENOYER-GEPPERT 112
    SOMMERVILLE HOUSE 164
    STANTON HOBBY 109
ANDY VODA OPTICAL TOYS 110, 155
ANEMOMETER 88
ANEROID BAROMETER 89
anesthesiology
    WOOD LIBRARY 76
animal models
    EDUCATIONAL TOYS 159
ant farms
    CHILDCRAFT EDUCATION 111
    UNCLE MILTON IND 165
anthropology
    NORTHERN ILLINOIS UNIVERSITY ANTHROPOLOGY MUSEUM 73
    ORIENTAL INSTITUTE MUSEUM 74
Apple computers
    EDUTAINMENT CATALOG 38
    MAC-USER 29
    MACWORLD 29
APPLE DIRECT 37
APPRAISAL: SCIENCE BOOKS FOR YOUNG PEOPLE 123
ARBOR SCIENTIFIC 111
arboretums
    EVANSTON ECOLOGY CENTER 67
    EVENTS, NEWS & CLASSES 136
    MORTON ARBORETUM 72
    STERLING MORTON LIBRARY 99
ARCHAEOLOGY 133
ARCHIE MC PHEE 111, 155
Argonne National Laboratory
    A CLASS ACT 127
    NEWTON 34
    SCIENCE EXPLORERS PROGRAM 53
    Special events 55
    TECHNICAL INFORMATION SERV 94
    U. S. DEPARTMENT 87
artesian well
    PILCHER PARK 74
arts
    COLUMBIA COLLEGE 47
    LIGHT WAVE 108
    MUSEUM OF HOLOGRAPHY 69

    SCHOOL OF HOLOGRAPHY 48
ARVEY PAPER & OFFICE PRODUCTS 105
ASIMOV'S BIOGR ENCYCLOPEDIA 14
ASIMOV'S CHRONOLOGY OF SCIENCE 14
assessment
    AURBACH & ASSOCIATES 33
    SYNAPSYS SOFTWARE 35
ASSOCIATION FOR WOMEN IN SCIENCE 85
astronomy
    ABRAMS SKY CALENDAR 123
    ADLER PLANETARIUM 44, 65
    ADLER PLANETARIUM SHOP 104
    ASTRONOMY FOR EVERY KID 8
    ASTRONOMY MAGAZINE 133
    CERNAN EARTH AND SPACE CTR 65
    CHICAGO ASTRON 78
    COSMIC QUARTERLY 134
    DEEP SKY 134
    KNOLLWOOD BOOKS 25
    LITTLE RED SCHOOL 72
    ODYSSEY 124
    PROJECT STAR 119
    SKY & TELESCOPE 140
    TELESCOPE MAKING 140
    TIME MUSEUM 76
    TIME MUSEUM STORE 23
    TOY STATION 109
ASTRONOMY FOR EVERY KID 8
ASTRONOMY MAGAZINE 133
AUDUBON 133
AURBACH & ASSOCIATES 33
awards - local 55
awards - national 60
B DALTON BOOKSELLER 19
BABYBUG 123
BALANCE 89
BAREBONES 106, 152
BARNARD COMPUTER, MATHEMATICS & SCIENCE CENTER 42
BARNES & NOBLE BOOKSTORES 20
BAROGRAPH 89
BENCHMARKS FOR SCIENCE LITERACY PROJECT 2061 16
BEST BOOKS FOR CHILDREN 2
BEST BUY 35
BEST REF BOOKS FOR YOUNG PEOPLE 2
BEST'S SAFETY DIRECTORY 141
BILL NYE THE SCIENCE GUY 169
BIO WEST 78

Index 177

biology
    ADAM SOFTWARE 32
    AMERICAN BIOLOGY TEACHER 127
    BIO WEST 78
    BIOLOGY SURVIVAL GUIDE 14
    BIOSCIENCE 133
    CAROLINA BIOL 111
    CAROLINA TIPS 127
    CLASSROOM CREATURE 8
    DENOYER-GEPPERT 112
    ILLINOIS ASSOC 80
    LIFE SCIENCES 136
    NATL ASSOC OF BIOL TEACH 25, 86
    NEBRASKA SCIENTIFIC 118
    THREE RIVERS AMPHIBIAN 120
    WARD'S 121
    WISCONSIN FAST PLANTS 121
    YOUNG ENTOMOLOGIST'S SOC 121
BIOLOGY TEACHER'S SURVIVAL GUIDE 14
BIOSCIENCE 133
bird watching
    CHICAGO BOTANIC GARDEN 46
    DAMERT COMPANY 158
    HIKING & BIKING 64
    KLUTZ 161
    NATURE COMPANY 109
    NORTH PARK VILLAGE 73
BLOCKBUSTER VIDEO 166
BLOCKS AND MARBLES BRAND TOYS 156
BOARDWATCH MAGAZINE 27
BOGAN COMPUTER TECHNICAL H S 42
book loan
    SCIENCE 2001 TEXT SETS 6
BOOK LOVER'S GUIDE TO CHGOLAND 19
book reviews
    APPRAISAL 123
    SCIENCE BOOKS & FILMS 138, 168
    SCITECH BOOK NEWS 139
BOOK STALL OF ROCKFORD 20
BOOKMAN'S CORNER 20
books - mail order 23
BOOKS & AV MATERIALS FOR CHILDREN 2
BOOKSELLERS ROW 20
bookstores - local 19
BOOKWORKS 20
BORDERS BOOKS 20
botany
    N. FAGIN BOOKS 22

bridge building
    BRIDGE BUILDING CONTEST 57
    MIDWEST MODEL 108
    MIDWEST PRODUCTS 117
BRODERBUND SOFTWARE 38
BROOKFIELD ZOO 65
    EDUC DEPT 45
    LIBRARY 94
    PREVIEW 133
    SPECIAL EVENTS 57
BROOKFIELD ZOO - BOOKSTORE 20
BROOKFIELD ZOO PREVIEW 133
bugs
    YOUNG ENTOMOLOGIST'S SOC 121
bulletin boards
    BOARDWATCH MAGAZINE 27
    NASA SCIENCE BBS 32
    NEWTON 34
    PRODIGY 34
    PROJECT INFORM 34
    STIS 30
CAMP SAGAWAU 65
CAPRICORN TOYS, INC. 156
Carmen Sandiego
    BRODERBUND 38
CAROLINA BIOLOGICAL SUPPLY 111
CAROLINA TIPS 127
CASPAR 45
CD-ROM
    ADVENTURES OF HYPERMAN 169
    CD-ROM WORLD 28
    DAVIDSON & ASSOC 38
    EDUCATIONAL ACTIV 38
    EDUTAINMENT CATALOG 38
    SCHOLASTIC SOFTW 39
    SCIENCE HELPER K-8 35
    SOFTWARE PLUS 39
CD-ROM WORLD 28, 134
CENCO 111
CENCO REFLECTIONS 134
CENTER FOR ADVANCED SPACE STUDIES 32
CENTER FOR RESEARCH LIBRARIES 94
CERNAN EARTH AND SPACE CENTER 65
CHALLENGE OF THE UNKNOWN 170
CHANNEL 2 NEWS WEATHER TEAM 45
CHEM MATTERS 127
CHEM WEST 78
CHEMICAL EDUCATION 127
CHEMICAL SAFETY 141

178 Science Fun in Chicagoland

chemistry
    AM CHEMICAL SOCIETY - CHGO 78
    AMERICAN CHEMICAL SOCIETY 84
    CHEM MATTERS 127
    CHEM WEST 78
    CHEMICAL EDUCATION 127
    CHEMICAL SAFETY 141
    CHEMUNITY NEWS 127
    FLINN CHEMICAL 115
    ICE PICKS 3
    KITCHEN SCIENCE 11
    SCHOLARSHIP EXAMINATION 55
    STANTON HOBBY 109
    TEACHING RESOURCES CAT 110
    TODAY'S CHEMIST 140
    TROST HOBBY SHOP 154
    TYCO TOYS 165
    VERNIER SOFTWARE 39
CHEMUNITY NEWS 127
CHESTER ELECTRONICS SUPPLY 106
CHICAGO ACADEMY OF SCIENCES 65
    EDUC DEPT 45
    LIBRARY 95
    NATURE'S NOTES 137
CHICAGO ASTRONOMICAL SOCIETY 78
CHICAGO BOTANIC GARDEN 66
    LIBRARY 95
    REGISTRAR'S OFFICE 46
    THE GARDEN SHOP 106
CHICAGO CHILDREN'S MUSEUM 46, 66
    ENVIROMANIA 57
    INVENTING FAIR 57
    RECYCLED RALLY 57
CHICAGO COMPUTER EXCHANGE 35
CHICAGO KITE CO. 106, 152
CHICAGO NON-PUBLIC SCHOOLS ANNUAL SCI EXPOSITION 148
CHICAGO PUBLIC LIBRARY 95
CHICAGO PUBLIC SCHOOLS - CURRICULUM GUIDES 15
CHICAGO PUBLIC SCHOOLS ANNUAL STUDENT SCIENCE FAIR 148
CHICAGO ROCKS & MINERALS SOCIETY 78
CHICAGO STATE UNIV
    LIBRARY 96
CHICAGO SYSTEMIC INITIATIVE 46
CHICAGO ZOOLOGICAL SOCIETY
    LIBRARY 96
CHICAGO'S MUSEUMS 63
CHGOLAND SKY LINERS KITE CLUB 57, 79
CHICKADEE 124

CHILDCRAFT EDUCATION CORP. 111, 156
children's books
    CHILDREN'S BOOKSTORE 21
    LITTLE MINDS 25
CHILDREN'S BOOKSTORE 21
CIRCUIT CITY 35
CITY COLLEGES OF CHICAGO 28
CLASS MATE 106
CLASSROOM CREATURE CULTURE 8
CLEARBROOK CENTER - LEKOTEK 151
clubs
    AMERICAN RADIO RELAY LEAGUE 44
COLE-PARMER INSTRUMENT CO. 111
COLLEGE VIDEO 166
COLUMBIA COLLEGE CHICAGO
    INSTITUTE FOR SCI EDUC 47
COMMONWEALTH EDISON COMPANY
    LIBRARY 96
community
    SCIENCE LINKAGES IN THE COMMUNITY 53
competitions - local 55
competitions - national 60
COMPRESSED AIR MAGAZINE 128
CompUSA 35
COMPUTE 28, 134
COMPUTER CHRONICLES 28
COMPUTER DISCOUNT WAREHOUSE 36
computer educator discounts
    APPLE DIRECT 37
    SOFTWARE PLUS 39
    SOFTWARE SOURCE 39
computer information sources 27
computer suppliers - local 35
computer suppliers - mail order 37
computer training
    CITY COLLEGES OF CHICAGO 28
    COMPUTER CHRONICLES 28
    TEACHERS ACADEMY 54
COMPUTERLAND 36
computers in the classroom 32
    AURBACH & ASSOCIATES 33
    COMPUTING TEACHER 28
    I.C.E. CUBE 29
    IL COMPUTING EDUCATORS 33
    ILLINOIS COMP 80
    INTERNATIONAL SOC 85
    INTERNL SOCIETY FOR TECHN 38
    LAP TOP COMPUTER EDUC 33
    MEDIA AND METHODS MAG 29
    PRODIGY 34

SYNAPSYS SOFTWARE 35
T.H.E. JOURNAL 30
TECHN. AND LEARNING MAG 30
TECHNOLOGY TEACHER 30
COMPUTING ENGINEERS INC. 31
COMPUTING TEACHER 28, 128
CONSERVATION EDUCATION CATALOG 1
CONSTRUCTIVE PLAYTHINGS 106, 111, 152, 156
cooperative learning
    NASA/NSTA SPACE 61
    SCIENCE EXPERIENCES 12
COPERNICUS 156
COSMIC QUARTERLY 134
courses
    ADLER PLANETARIUM 44
    ARGONNE NATIONAL LAB 44
    CAMP SAGAWAU 65
    CHICAGO ACADEMY OF SCI 45
    CHICAGO BOTANIC GARDEN 46
    FIELD MUSEUM 47
    JOHN G. SHEDD AQUARIUM 49
    LEDERMAN SCI EDUC CENTER 51, 71
    MORTON ARBORETUM 51
    TEACHERS ACADEMY 54
    WALTER E. HELLER 76
CRABTREE NATURE CENTER 66
CRAFT HOUSE CORP. 156
CRC HANDBOOK OF CHEMISTRY AND PHYSICS 15
CREATIVE PUBLICATIONS 112, 157
CREATIVITY FOR KIDS 157
CRERAR LIBRARY 101
CRITICAL THINKING PRESS & SOFTW 24
CUISENAIRE CO. OF AMERICA 112
CURIOSITY KITS 157
CURIOSITY PLACE 73
CURRENT SCIENCE 124, 128
curricula
    CHICAGO PUBLIC SCHOOLS 15
    DELTA EDUCATION 112
    NATIONAL DIFFUSION NETWORK 41
    NATL SCI RESOURCES CENTER 41
    PROGRAMS THAT WORK 16
    PROMISING PRACTICES 16
    SCIENCE FOR CHILDREN 6
    SCIENCE HELPER 7
CUT RATE TOYS 106, 153
DALE SEYMOUR PUBLICATIONS 24, 112
DAMERT COMPANY 158

DAN BEHNKE, BOOKSELLER 21
DAVE'S ROCK SHOP 106
DAVIDSON & ASSOCIATES 38
DE PAUL UNIVERSITY
    LIBRARY 96
DE VRY INSTITUTE OF TECHNOLOGY 96
DEEP SKY 134
DELTA EDUCATION, INC. 112, 158
DEMCO 38
DEMO EXPERMTS IN PHYSICS 9
DEMONSTRATION HANDBOOK FOR PHYSICS 9
DENOYER-GEPPERT 112
DES PLAINES VALLEY GEOLOGICAL SOCIETY 79
DESIGN SCIENCE TOYS LTD. 158
DEVON TOYS 106, 153
DICTIONARY OF SCIENTIFIC BIOGRAPHY 15
DIDAX INC. EDUCATIONAL RESOURCES 112
DINOSAUR TIMES 126
dinosaurs
    DINOSAUR TIMES 126
    EDUCATIONAL TOYS 114, 159
    GALT COMPANY, INC. 160
    INVESTIGATING SCIENCE 11
    PLAY VISIONS 163
    STANTON HOBBY 109
    TAMIYA AMERICA, INC. 164
    TYCO TOYS 165
DIRECTORY OF SPECIAL LIBRARIES AND INFORMATION CENTERS 92
disabilities
    CLEARBROOK CENTER 151
    SAVI/SELPH 119
    SCIENCE ASSOCIATION FOR 139
    TEACHING CHEMISTY 18
DISCOVER 134
DISCOVER OUR WORLD RESOURCE BOOK 2
DISCOVERY CENTER 112
DISCOVERY CENTER MUSEUM 47, 66
DNA research
    SECRET OF LIFE 172
DOLPHIN LOG 124
DOOLIN AMUSEMENT SUPPLY CO. 107, 153
DOWLING MINER MAGNETICS CORP. 113
DRIVER'S POCKET GUIDE TO HAZARDOUS MATERIALS 142
DUPAGE CHILDREN'S MUSEUM 66
DUPONT CHALLENGE 60
DURACELL/NSTA SCHOLARSHIP COMPETITION 60

180 Science Fun in Chicagoland

EARLY CHILDHOOD AND SCIENCE 9
earth science 84
    AM GEOLOGICAL INSTITUTE 84
    SCOTT RESOURCES 120
EARTH SCIENCE CLUB OF NORTHERN ILLINOIS 79
EARTH SCIENCE FOR EVERY KID 9
EARTH-FRIENDLY TOYS 150
EARTHDWELLER 107
ECHO SOUNDER 89
EDGEWATER, UPTOWN, ROGERS PARK SCIENCE CLUBS 79
EDMUND SCIENTIFIC COMPANY 113, 158
EDTALK: WHAT WE KNOW ABOUT MATH 15
EDTALK: WHAT WE KNOW ABOUT SCIENCE 15
education opportunities 44
EDUCATIONAL ACTIVITIES, INC. 38, 167
EDUCATIONAL DESIGN, INC. 158
EDUCATIONAL INSIGHTS 159
EDUCATIONAL RESOURCES 38
EDUCATIONAL TEACHING AIDS 107
EDUCATIONAL TOYS, INC. 114, 159
EDUCATORS GUIDE TO FREE COMPUTER MATERIALS 28
EDUCATORS GUIDE TO FREE SCIENCE MATERIALS 103
EDUCORP 38
EDUTAINMENT CATALOG 38
EGG HEAD SOFTWARE 36
EISENHOWER MATHEMATICS AND SCIENCE 40, 82
EISENHOWER NATIONAL CLEARINGHOUSE 3, 34
electricity
    COMMONWEALTH EDISON 96
    POWER HOUSE 74
    POWER HOUSE CATALOG 119
    POWER HOUSE RESOURCE CTR 100
ELECTRONIC LEARNING 128
electronics
    CHESTER ELECTRONICS 106
    MOTOROLA MUSEUM 72
    SYLVESTER ELECTRICAL 109
ELEKTEK 37
elementary particle physics
    FERMILAB LIBRARY 101
elementary school
    SCIENCE FOR THE 17
    SCIENCE IN 17
    TEACHING SCIENCE THROUGH 18

ELEM SCHOOL SCIENCE FOR THE 90'S 15
ELGIN PUBLIC MUSEUM 67
ELGIN ROCK & MINERAL SOCIETY 79
ENCHANTED KINGDOM 153
energy
    ENERGY EDUC 104
    IL DEPT OF ENERGY LIBR 97
    NATIONAL ENERGY FOUNDATION 86
    U. S. DEPARTMENT OF 87
ENERGY EDUCATION RESOURCES 3, 104
engineering
    BRIDGE BUILDING CONTEST 57
    CHGOLAND KITE CLUB 57
    GALVIN LIBRARY 98
    IIT 100 SPEEDWAY 58
    IIT INDUSTRIAL DESIGN 58
    ILLINOIS JETS 80
    ILLINOIS SCIENCE OLYMPIAD 80
    JETS 86
    JETS ENGINEERING 58
    LEGO DACTA 117
    NATIONAL ACTION 86
    OWI INCORPORATED 163
    TAMIYA AMERICA, INC. 164
entomology
    UNCLE MILTON IND 165
    YOUNG ENTOMOLOGIST'S SOC 121
environment
    ALLIANCE FOR ENVIRON 84
    ARGONNE NATL LABORATORY 45
    CHICAGO BOTANIC GARDEN 46
    CHGO CHILDREN'S MUSEUM 46, 57
    COMMONWEALTH EDISON 96
    CONSERVATION EDUC CATALOG 1
    EARTH-FRIENDLY TOYS 150
    ENVIRONMENTAL EDUC 79
    EVANSTON ECOLOGY CENTER 21, 67
    FOREST PRESRVE DUPAGE CO 48
    FORESTRY SUPPLIERS 115
    FULLERSBURG WOODS 68
    FUTURES 135
    ICEWALK 170
    ILLINOIS AUDUBON SOC 80
    INSTITUTE OF ENVIR 85
    JOURNAL OF INST OF ENVIRON 136
    LAKE CO FOREST PRESRVES 50, 71
    PLASTICS AND THE ENVIR 104
    PLASTICS IN OUR WORLD 119
    RACE TO SAVE THE PLANET 171
    SPRING VALLEY 54
    U. S. ENVIR LIBRARY 101

ENVIRONMENTAL EDUCATION RESOURCE LIST 64
ENVIRONMENTAL EDUCATION ASSOCIATION OF ILLINOIS 79
ENVIRONMENTAL EDUCATION RESOURCE LIST 3
equipment 103
equipment - mail order 110
ESS curriculum materials
    DELTA EDUCATION 112
essay contest
    DUPONT CHALLENGE 60
ESTES INDUSTRIES 114, 159
ETA 114
EUREKA! 24, 114
EVANSTON ECOLOGY CENTER 67
EVANSTON ECOLOGY CTR - BOOKSTORE 21
events - local 55
events - national 60
EVERY TEACHER'S SCIENCE BOOKLIST 3
EVERYDAY SCIENCE SOURCEBOOK 9
excursions 63
EXPERIMENTAL SCIENCE 9
EXPERIMENTING WITH INVENTIONS 10
EXPERIMENTING WITH MODEL ROCKETS 10
EXPLORATORIUM SCIENCE SNACKBOOK 10
EXPLORATORIUM STORE 24, 114, 159
EXPLORATOY 159
EXPLORING SCIENCE 3
EXTRAPOLATOR 128
FABYAN MUSEUM 67
FACTS ON FILE SCIENTIFIC YEARBOOK 134
FEDERAL GOVERNMENT DATABASES 32
FERMILAB
    GUIDED TOURS 67
    LIBRARY 101
    SCIENCELINES 132
FIELD MUSEUM 68
    BOOKSHOP 21
    HARRIS LOAN PROGRAM 114
    LIBRARY 96
field trips 63
    AMUSEMENT PARK PHYSICS 8
    BROOKFIELD ZOO 45
    CAMP SAGAWAU 65
    CHICAGO ACADEMY OF SCIENCES 65
    CHICAGO BOTANIC GARDEN 46, 66
    CHICAGO CHILDREN'S MUSEUM 66
    CRABTREE NATURE CENTER 66
    DISCOVER OUR WORLD 2

DISCOVERY CENTER MUSEUM 66
ENVIRONMENTAL EDUCATION 3
FERMILAB 67
FIELD MUSEUM 47, 68
FOREST PRESERVE DUPAGE CO 48
FULLERSBURG WOODS 68
GARFIELD PARK CONSERVATORY 68
JOHN G. SHEDD AQUARIUM 49
LAKE CO FOREST PRESERVES 50
PILCHER PARK 74
POWER HOUSE 74
RESOURCES FOR THE CLASSROOM 6
RYERSON CONSERVATION AREA 67
SAND RIDGE 75
SCITECH 75
SIX FLAGS 60, 75
SPRING VALLEY 75
TEACH THE MIND 13, 64
TEN-MINUTE FIELD TRIPS 13, 64
TREKS FOR TROOPS 64
film reviews
    SCIENCE BOOKS & FILMS 168
FISHER-EMD 115
FLINN CHEM CAT REF MANUAL 115, 144
FLYING APPARATUS CATALOGUE
    KLUTZ 161
forest preserve districts
    CAMP SAGAWAU 65
    COOK COUNTY 48
    CRABTREE NATURE CENTER 66
    DUPAGE COUNTY 48
    FULLERSBURG WOODS 68
    KLINE CREEK FARM 70
    LAKE COUNTY 50, 71
    LITTLE RED SCHOOL 72
    RIVER TRAIL NATURE CENTER 74
    RYERSON CONSERVATION AREA 67
    SAND RIDGE NATURE CENTER 75
    SPRING VALLEY 75
    TRAILSIDE MUSEUM 76
    WALTER E. HELLER 76
    WILLOWBROOK 76
FORESTRY SUPPLIERS, INC. 115
fossils
    JURICA NATURE MUSEUM 70
    PREHISTORIC LIFE MUSEUM 74
free computer materials
    EDUCATORS GUIDE 28
free films
    EDUCATORS GUIDE 167

182 Science Fun in Chicagoland

free materials
    EDUCATORS GUIDE 2, 167
free videos
    EDUCATORS GUIDE 167
FREY SCIENTIFIC 115
frogs and tadpoles
    THREE RIVERS AMPHIBIAN 120
FULLERSBURG WOODS ENVIRONMENTAL EDUCATION CENTER 68
FUTURE SCIENTISTS & ENGINEERS OF AMERICA 85
FUTURES 135
FUTURES-1 170
FUTURES-2 170
GALILEO SCHOLASTIC ACADEMY OF MATH AND SCIENCE 42
GALT COMPANY, INC. 160
GALT TOYS 153
GALVIN LIBRARY 98
GARDEN TALK 135
GARFIELD PARK CONSERVATORY 68
GAS RESEARCH INSTITUTE
    LIBRARY 96
GENERAL SCIENCE INDEX 122
GEOCENTER 115
geology
    AM GEOL INSTITUTE 24, 84
    CHICAGO ROCKS & MIN 78
    DAVE'S ROCK SHOP 106
    DES PLAINES VALLEY GEOL 79
    EARTH SCIENCE CLUB 79
    EARTH SCIENCE FOR 9
    ELGIN PUBLIC MUSEUM 67
    ELGIN ROCK & 79
    GEOCENTER 115
    GEOTIMES 128
    GREATER OAKLAWN 79
    ISGS GEONEWS 135
    JOURNAL OF GEOLOGICAL EDUC 128
    LAKE COUNTY GEM 82
    LIZZADRO MUSEUM 72, 108
    MIDWEST HISTORICAL RES 83
    NATIONAL ENERGY 86
    NATL ASSOC OF GEOL TEACHERS 86
    NATL ENERGY FOUNDATION 25
    PARK FOREST 83
    PREHISTORIC LIFE MUSEUM 74
    RESOURCES FOR TEACH GEOLOGY 4
    THE GARDEN SHOP 106
    WEST SUB LAPIDARY 83
GEOSPACE PRODUCTS COMPANY 159

GEOTIMES 128
GERSHWIN MATH AND SCIENCE COMMUNITY ACADEMY 42
gifted
    EARLY CHILDHOOD 9
    SCIENCE & ARTS ACADEMY 43
GLOBAL LAND INFORMATION SYSTEMS 32
globes
    CHILDCRAFT EDUCATION 111
    PLAY VISIONS 163
GOLDEN APPLE FOUNDATION 58
GOLDEN APPLE SCIENCE PROGRAM 48
GOVERNORS STATE UNIVERSITY
    LIBRARY 97
GRADUATED CYLINDER 89
GRAND VALLEY STATE UNIVERSITY SCIENCE & MATHEMATICS UPDATE 135
GRAY'S DISTRIBUTING CO., INC. 107
GREAT EXPECTATIONS 21
GREATER CHICAGO SAFETY COUNCIL 142
GREATER OAKLAWN DIGGERS 79
groups - local 78
groups - national 84
GROVE 68
guest speakers
    CHANNEL 2 WEATHER TEAM 45
    CHICAGO CHILDREN'S MUSEUM 46
    DISCOVERY CENTER 47, 66
    FOREST PRESERVE COOK CO 48
    ILLINOIS DEPT OF ENERGY 48
    POWER HOUSE 52, 60
    SPRING VALLEY 54
    WIERD SCIENCE KIDS 60
GUIDE TO FREE FILMS 167
GUIDE TO FREE SCIENCE MATERIALS 167
GUIDE TO FREE VIDEOTAPES 167
GUIDE TO MATH & SCI REFORM 18, 33
HAMILL & BARKER 21
hands-on materials 103
hands-on materials - mail order 110
HAROLD WASHINGTON LIBRARY CENTER SCIENCE FAIRS 146
HARRIS EDUCATIONAL LOAN PROGRAM 114
HARRISON SUPPLY 107
HAWKHILL ASSOCIATES 167
HAZARDOUS MATERIALS 142
HAZARDOUS WASTE RESEARCH AND INFORMATION CENTER 143
HELENA SZEPE, BOOKS 21
HELPING YOUR CHILD LEARN SCIENCE 10

HIKING & BIKING IN LAKE CO, IL 64
history
    ASIMOV'S BIOGR. ENCYCLOPEDIA 14
    ASIMOV'S CHRONOLOGY 14
    CRERAR LIBRARY 101
    DICTIONARY OF BIOGRAPHY 15
    HAMILL & BARKER 21
    HELENA SZEPE, BOOKS 21
    HISTORY OF SCI SOC 85
    ISIS 136
    RICHARD ADAMIAK 22
    SMITHSONIAN 140
    ST. JAMES'S HOUSE 109
    TIMETABLES OF SCIENCE 18
    TRADITION OF SCIENCE 18
    U OF CHGO SPECIAL COLL 102
HISTORY OF SCIENCE SOCIETY 85
holography
    LIGHT WAVE 108
    MUSEUM OF HOLOGRAPHY 69
    SCHOOL OF 48
HOMECRAFTERS MANUFACTURING 160
HONEYWELL, INC.
    LIBRARY 97
horology
    TIME MUSEUM 76
HORTICULTURAL SALES PRODUCTS 160
horticulture
    CHICAGO BOTANIC GARDEN 46, 66, 95
    FORESTRY SUPPLIERS 115
    GARDEN TALK 135
    GARFIELD PARK CONSERVATORY 68
    HORTICULTURAL SALES PRODUCTS 160
    HORTICULTURE 135
    KLINE CREEK FARM 70
    LIBBY LEE TOYS 161
    LILACIA PARK 71
    LINCOLN PARK CONSERVATORY 71
    LIVING CLASSROOMS 117
    OAK PARK CONSERVATORY 73
    PILCHER PARK 74
    Root-Vue-Farm 160
    THE GARDEN SHOP 106
    WISCONSIN FAST PLANTS 121
HUBBLE TELESCOPE NEWS 32
HYDROMETER 89
HYGROMETER 89
I.C.E. CUBE 29, 33, 135
ICE PICKS 3

ICEWALK 170
IDEA FACTORY, INC. 24, 116
IDEAAAS 4
IDEAL SCHOOL SUPPLY CO 116, 160
IIT 100 SPEEDWAY 58
IIT INDUSTRIAL DESIGN OPEN HOUSE 58
ILLINOIS ASSOC OF BIOLOGY TEACHERS 80
ILLINOIS AUDUBON SOCIETY 80
ILLINOIS COMPUTING EDUCATORS 80
ILLINOIS DEPARTMENT OF ENERGY 48
ILLINOIS DEPT OF ENERGY
    LIBRARY 97
ILLINOIS INSTITUTE OF TECHNOLOGY
    LIBRARY 98
    MANUF TECHN CENTER 98
    BOOKSHOP 21
ILLINOIS JETS 80
ILLINOIS JUNIOR ACADEMY OF SCI 148
ILLINOIS MATH AND SCIENCE ACADEMY 42
ILLINOIS SCIENCE OLYMPIAD 58, 80
ILLINOIS SCIENCE TEACHERS ASSOC 80
ILLINOIS STATE ACADEMY OF SCIENCE
    TRANSACTIONS 135
ILLINOIS STATE BOARD OF EDUCATION 40
    INTERNET ACCESS 31
ILLINOIS STATE GEOLOGICAL SURVEY 4
ILLINOIS STATE PHYSICS PROJECT 81, 82
INFORMATION SOURCES IN SCIENCE 4
insects
    EXPLORATOY 159
    UNCLE MILTON IND 165
    YOUNG ENTOMOLOGIST'S SOC 121
INSIGHT MEDIA 167
INSTITUTE FOR MATH AND SCI EDUC 49
INSTITUTE OF ENVIRONMENTAL SCI 85
INSTITUTE OF GAS TECHNOLOGY
    TECHN INFO CENTER 98
INTERACTIVE PHYSICS 33
interactive television
    SCIENCE POWER 171
INTERNATIONAL MUSEUM OF SURGICAL SCIENCE 70
INTERNATIONAL SCIENCE AND ENGINEERING FAIR 61, 148
INTERNATIONAL SOCIETY FOR TECHNOLOGY IN EDUCATION 85
INTERNATIONAL TECHNOLOGY EDUCATION ASSOCIATION 24, 85

184 Science Fun in Chicagoland

Internet 31
    AIR POLUTION 32
    BOARDWATCH MAGAZINE 27
    COMPUTING ENGINEERS 31
    CTR FOR ADV SPACE STUDIES 32
    FED GOVERNMNT DATABASES 32
    GLOBAL LAND INFO 32
    HUBBLE TELESCOPE NEWS 32
    IL STATE BOARD OF EDUC 31
    NASA SCIENCE BBS 32
    NEWTON 32, 34
    OCEANOGR & ATMOSPH ADMIN 32
    PRODIGY 34
    SCI AND TECHNOLOGY INFO 32
INTERNL SOCIETY FOR TECHNOLOGY IN EDUCATION 38
inventing
    AMERICAN SCI & SURPLUS 104
    CHICAGO CHILDREN'S MUSEUM 57
    DURACELL/NSTA 60
    EXPERIMENTING WITH 10
    HARRISON SUPPLY 107
    INVENTING, INVENTIONS 10
    INVENTOR'S COUNCIL 49
    KOLBE CONCEPTS 116
    MIDWEST MODEL 108
    TOSHIBA NSTA 62
INVENTING THE FUTURE 167
INVENTING, INVENTIONS & INVENTORS 10
INVENTOR'S COUNCIL 49, 82
INVESTIGATING SCI WITH DINOSAURS 11
INVITATIONS TO SCIENCE INQUIRY 10
ISGS GEONEWS 135
ISIS 136
J. WESTON WALCH, PUBLISHER 24
J.C.'S KITES 107, 153
JETS ENGINRING DESIGN COMPETITION 58
JOHN CRERAR LIBRARY 101
JOHN G. SHEDD AQUARIUM 49, 70
    LIBRARY 98
JOURNAL OF COLLEGE SCI TEACHING 129
JOURNAL OF GEOLOGICAL EDUCATION 128
JOURNAL OF RES IN SCI TEACHING 129
JOURNAL OF SCI TEACHER EDUCATION 131
JOURNAL OF THE INSTITUTE OF ENVIRONMENTAL SCI 136
journals 122
journals on science 133
JUNIOR ENGINEERING TECHN SOCIETY 86

junior high school science
    SCIENCE SCOPE 131
    MID-LEVEL NETWORK 82
JURICA NATURE MUSEUM 49, 70
KADON ENTERPRISES, INC. 160
KAY BEE TOYS 107, 153
KEY CURRICULUM PRESS 25
KIPP BROTHERS, INC. 116, 160
KITCHEN SCIENCE 11
KITE HARBOR 107, 153
KITELINES 124
kiting
    AMERICAN KITE 123
    CHICAGO KITE CO. 106
    CHGOLAND SKY LINERS KITE CLUB 57, 79
    J.C.'S KITES 107
    KITE HARBOR 107
    KITELINES 124
    TROST HOBBY SHOP 154
kits
    CREATIVITY FOR KIDS 157
    CURIOSITY KITS 157
    DELTA EDUCATION 158
    EDUCATIONAL DESIGN 158
    HOMECRAFTERS MANUF 160
    IDEAL SCHOOL SUPPLY 160
    KLUTZ 161
    OWI INCORPORATED 163
    SMALL WORLD TOYS 164
KLINE CREEK FARM 70
KLUTZ 116, 161
KNOLLWOOD BOOKS 25
KOHL CHILDREN'S MUSEUM 70
KOHL LEARNING STORE 107, 153
KOLBE CONCEPTS, INC. 116, 161
KRASNY & COMPANY INC. 108
KROCH'S & BRENTANO'S INC. 22
LAB SAFETY SUPPLY INC. 116, 144
laboratory equipment 103
laboratory equipment - mail order 110
LAKE COUNTY EDUCATIONAL SERVICE CENTER 50
LAKE COUNTY FOREST PRESERVES 68, 71
LAKE COUNTY GEM & MINERAL SOCIETY 82
LAKESHORE LEARNING MATERIALS 117, 161
LANE TECHNICAL HIGH SCHOOL 42
language
    QUICK SCIENTIFIC TERMINOLOGY 12
LAP TOP COMPUTER EDUCATION 33

laserdisc sales
    ZTEK CO. 169
Lawrence Hall of Science
    DISCOVERY CENTER 112
    EUREKA! 24, 114
LEARNING RESOURCES, INC. 117, 161
LEARNING SERVICES 38
LEDERMAN SCI EDUC CTR 34, 51
    INSTRUCTIONAL MATERIALS 108
    SCIENCELINES 132
    TEACHER RESOURCE CENTER 98
LEGO DACTA 117, 161
lesson plans
    SCIENCE HELPER 7, 35
LIBBY LEE TOYS, INC. 161
libraries 92
LIBRARIES UNLIMITED 25
LIFE SCIENCES 136
LIGHT WAVE 108, 154
LILACIA PARK 71
LINCOLN PARK CONSERVATORY 71
LINCOLN PARK ZOOLOGICAL GARDENS 71
    LIBRARY 98
Lincoln Park Zoological Soc
    ZOO REVIEW 140
LINCOLN PARK ZOOLOGICAL SOCIETY 82
LINDSAY PUBLICATIONS 9
literature
    BEST BOOKS FOR CHILDREN 2
    EVERY TEACHER'S SCI BOOKLIST 3
    SCIENCE IN FACT AND FICTION 6
    SCIENCE THROUGH CHILDREN'S 18
    SOMMERVILLE HOUSE 164
LITTLE MINDS: BOOKS FOR KIDS 25
LITTLE RED SCHOOL HSE NATURE CTR 72
LIVING CLASSROOMS 117
LIZZADRO MUSEUM OF LAPIDARY ART 72
    SHOP 108
LOYOLA UNIVERSITY CHICAGO
    SCIENCE LIBRARY 99
LOYOLA UNIV FAMILY SCIENCE NIGHT 59
MAC/CHICAGO 29
MAC-USER 29, 136
MACWORLD 29, 136
magazines 122
magazines for teachers 126
magazines on science 133
magnetism
    CHILDCRAFT EDUCATION 111
    DOWLING MINER MAGNETICS 113
    GEOSPACE PRODUCTS 159

Index 185

MANUFACTURING TECHNICAL INFORMATION
    ANALYSIS CENTER 98
MATH AND SCIENCE HANDS-ON 117
MATH SCI CONSORTIA DEMO SITE 34
MATH, SCIENCE, AND TECHNOLOGY
    PROGRAMS THAT WORK 16
mathematics
    AIMS 110
    CHALLENGE OF THE UNKNOWN 170
    CUISENAIRE CO. 112
    DALE SEYMOUR 112
    DIDAX INC 112
    EDTALK 15
    EXTRAPOLATOR 128
    FUTURES-1 170
    FUTURES-2 170
    GALT COMPANY, INC. 160
    KEY CURRICULUM PRESS 25
    NATL COUNCIL OF TEACH 25
    QUANTUM 125
    SECOND VOYAGE OF THE MIMI 172
    TIMS 121
    VOYAGE OF THE MIMI 172
MC GRAW-HILL ENCYCLOPEDIA 16
MCGRAW-HILL YEARBOOK OF SCIENCE AND
    TECHNOLOGY 136
measurements in science 88
MEDIA AND METHODS MAGAZINE 29
medicine
    INTERNATIONAL MUSEUM OF
        SURGICAL SCIENCE 70
    MUSEUM OF ANESTHESIOLOGY 76
MEI/MICRO CENTER 39
MERCURY BAROMETER 89
meteorology
    WEATHERWISE 126
METERSTICK 90
METROPOLITAN WATER RECLAM DISTR
    TECHN LIBRARY 99
MICROMETER 90
microscopes
    AMERICAN SCI & SURPLUS 104
    EDMUND SCIENTIFIC 113
    MORITEX 117
    SCOPE SHOPPE 109
    SELSI COMPANY 120
    TOY STATION 109
MID-LEVEL NETWORK 82

186 Science Fun in Chicagoland

middle school
    CURRENT SCIENCE 128
    MID-LEVEL NETWORK 82
    SCIENCE SCOPE 131
MIDWEST CONSORTIUM FOR MATHEMATICS AND SCIENCE EDUCATION 82
MIDWEST HISTORICAL RES SOCIETY 83
MIDWEST MODEL 108
MIDWEST PRODUCTS CO., INC. 117
MIDWEST VISUAL 39
minerals
    GEOCENTER 115
    LIZZADRO MUSEUM SHOP 108
minorities
    NATIONAL ACTION 86
model rockets
    ESTES INDUSTRIES 114, 159
    EXPERIMENTING WITH 10
    KNOLLWOOD BOOKS 25
    MIDWEST MODEL 108
    STANTON HOBBY 109
MOHS HARDNESS SCALE 90
MONTGOMERY WARD ELECTRIC AVENUE 37
MORITEX 117
MORTON ARBORETUM 51, 72
    EVENTS, NEWS & CLASSES 136
    LIBRARY 99
MORTON ARBORETUM - EVENTS, NEWS & CLASSES 136
MORTON ARBORETUM QUARTERLY 136
MOTOROLA MUSEUM OF ELECTR 51, 72
    EDUCATION PROGRAMS 51
MOTOROLA, INC. - CIG/GSS LIBRARY 99
MOTOROLA, INC. - COMMUNICATION SECTOR LIBRARY 99
MUDD LIBRARY FOR SCIENCE AND ENGINEERING 100
MULTICULTURALISM IN MATHEMATICS, SCIENCE 11
multimedia
    COMPUTE 28
    SUNBURST 39
MUSEUM NOTEBOOK 129
MUSEUM OF HOLOGRAPHY 69
MUSEUM OF SCIENCE & INDUSTRY 72
    BOOKSTORE & SHOP 22, 108, 154
    CURIOSITY PLACE 73
    EDUCATION DEPT 52
    FESTIVALS 59
    LIBRARY 99
    MUSEUM NOTEBOOK 129

    NASA TEACH RESOURCE CTR 100
    SCIENCE CLUB NETWORK 83
    SCIENCE FAIRS 148
    TEACHER OPEN HOUSE 59
museums 63
    ADLER PLANETARIUM 65
    CHGO ACADEMY OF SCIENCES 65
    CHICAGO CHILDREN'S MUSEUM 66
    DISCOVERY CENTER MUSEUM 66
    DUPAGE CHILDREN'S MUSEUM 66
    ELGIN PUBLIC MUSEUM 67
    EXPLORING SCIENCE 3, 64
    FABYAN MUSEUM 67
    FIELD MUSEUM 68
    GROVE 68
    INTERNATIONAL MUSEUM OF SURGICAL SCIENCE 70
    KOHL CHILDREN'S MUSEUM 70
    LEDERMAN SCIENCE EDUCATION CENTER 71
    LIZZADRO MUSEUM 72
    MUS OF SCIENCE & INDUSTRY 72
    NORTHERN IL UNIV 73
    ORIENTAL INSTITUTE MUSEUM 74
    POWER HOUSE 74
    RESOURCES FOR SCIENCE 4
    SCITECH 75
    SHEDD AQUARIUM 70
    TRAILSIDE MUSEUM 76
    WOOD LIBRARY 76
N. FAGIN BOOKS 22
NASA SCIENCE BBS 32
NASA TEACHER RESOURCE CENTER 52, 100
NASA/NSTA SPACE SCIENCE STUDENT INVOLVEMENT PROGRAM 61
NASCO 118
NATIONAL ACADEMY PRESS 25
NATL ACTION COUNCIL FOR MINORITIES IN ENGINEERING 86
NATL ASSOCIATION FOR SCIENCE, TECHN AND SOCIETY 86
NATIONAL ASSOC OF BIOL TEACHERS 86
NATIONAL ASSOCIATION OF GEOLOGY TEACHERS 86
NATIONAL AUDUBON SOC SPECIALS 170
NATIONAL DIFFUSION NETWORK 41
NATIONAL ENERGY FOUNDATION 25, 86
NATIONAL GEOGRAPHIC WORLD 124
NATIONAL LEKOTEK CENTER 152
NATIONAL SAFETY COUNCIL 143
    LIBRARY 100, 143

NATIONAL SCI EDUCATION STANDARDS 41
NATIONAL SCI RESOURCES CENTER 41, 87
NATIONAL SCI TEACHERS ASSOCIATION 87
NATL ASSOC OF BIOLOGY TEACHERS 25
NATL COUNCIL OF TEACHERS IN MATH 25
NATL OCEANOGR & ATMOSPH ADMIN 32
NATURAL ENQUIRER 137
natural history
    ELGIN PUBLIC MUSEUM 67
    FIELD MUSEUM 68
    FIELD MUSEUM LIBRARY 96
    NATIONAL GEOGR 124
    NATURAL HISTORY 137
    TRAILSIDE MUSEUM 76
NATURAL WONDERS 108, 154
nature
    CAMP SAGAWAU 65
    CHICAGO BOTANIC GARDEN 46
    CHICKADEE 124
    CONSERVATION EDUC CATALOG 1
    CRABTREE NATURE CENTER 66
    EARTHDWELLER 107
    FOREST PRESERVE COOK CO 48
    FOREST PRESERVE DUPAGE CO 48
    GROVE 68
    HIKING & BIKING 64
    ILLINOIS AUDUBON SOC 80
    JURICA NATURE MUSEUM 49, 70
    LAKE CO FOREST PRESERVES 50, 71
    LITTLE RED SCHOOL 72
    MORTON ARBORETUM 51, 72
    NATIONAL AUDUBON SPECIALS 170
    NATURE 137
    NATURE COMPANY 109
    NATURE COMPANY CAT 118
    NATURE CONNECTIONS PROJ 100
    NORTH PARK VILLAGE 73
    PILCHER PARK 74
    RIVER TRAIL NATURE CENTER 74
    RYERSON CONSERVATION AREA 67
    SAND RIDGE 75
    SPRING VALLEY 75
    WALTER E. HELLER 54, 76
    WARD'S 121
    WILLOWBROOK 76
    YOUR BIG BACKYARD 126
NATURE COMPANY 109
NATURE COMPANY CATALOG 118
NATURE CONNECTIONS PROJECT 100
NATURE'S NOTES 137

NATURE'S TOYLAND 162
NATUREPRINT PAPER PRODUCTS 163
NCREL 4, 34
    MIDWEST CONSORTIUM 40
NEBRASKA SCIENTIFIC 118
NEW EXPLORERS
    FIELD MUSEUM 115
    MUS OF SCI & INDUSTRY 52
NEW EXPLORERS, THE 170
NEW SCIENTIST 137
NEWBERRY MATHEMATICS AND SCIENCE ACADEMY 42
NEWTON 32, 34
NORTH CENTRAL REGIONAL EDUCATIONAL LABORATORY 4
NORTH PARK VILLAGE NATURE CENTER 73
NORTHEASTERN ILLINOIS UNIVERSITY LIBRARY 100
NORTHWESTERN UNIVERSITY
    SCI & ENGR LIBRARY 100
NOVA 171
novelties
    ACE-ACME 155
    AMERICAN SCI & SURPLUS 152
    ARCHIE MC PHEE 155
    CUT RATE TOYS 153
    DOOLIN AMUSEMENT 153
    KIPP BROTHERS 160
    ORIENTAL TRADING CO 163
    UNCLE FUN 155
    UNIVERSAL SPECIALTIES 165
NSRC NEWSLETTER 129
NSTA 87
NSTA MEMBERSHIP & PUBLICATIONS 26
NSTA REPORTS! 129
NSTA SCI EDUCATION SUPPLIERS 29, 104
NSTA SUPPLIERS 168
NSTA'S EDUCATIONAL HORIZONS 129
nuclear energy
    AMERICAN NUCLEAR SOC 94
    RE-ACTIONS 138
NUTS & BOLTS 147
O'GARA & WILSON BOOKSELLERS, LTD. 22
OAK PARK CONSERVATORY 73
OCCUPATIONAL SAFETY AND HEALTH ADMINISTRATION 143
oceanography
    DOLPHIN LOG 124
    JOHN G. SHEDD AQUARIUM 70
    SHEDD AQUARIUM LIBRARY 98
ODYSSEY 124

188 Science Fun in Chicagoland

OFFICE DEPOT 37
OFFICEMAX 37
OLYMPIAD
    ILLINOIS SCIENCE 58, 80
    THE SCIENCE 87
OMNI 137
OPTICAL SOCIETY OF CHICAGO 118
optics
    ANDY VODA 110
    ANDY VODA OPTICAL TOYS 155
    EDMUND SCIENTIFIC 113
    OPTICAL SOCIETY OF CHICAGO 118
    PLAY VISIONS 163
    SELSI COMPANY 120
OREGON SCIENCE TEACHER 129
organizations 77
    RESOURCES FOR SCIENCE 4
organizations - local 78
organizations - national 84
ORIENTAL INSTITUTE MUSEUM 74
    EDUC DEPT 52
ORIENTAL TRADING CO., INC. 118, 163
ORYX 26
OWI INCORPORATED 163
OWL 124
parents
    ACCESS 2000 44, 78
    ADLER PLANETARIUM 44
    CASPAR 45
    CHILDREN'S BOOKSTORE 21
    HELPING YOUR CHILD 10
    LITTLE MINDS 25
    PUBLICATIONS FOR PARENTS 26
    SCIENCE FARE 16
    SCIENCE-BY-MAIL 53
    TEACH YOUR CHILD SCI 13
    TEACHING SCIENCE WITH TOYS 151
PARK FOREST EARTH SCIENCE CLUB 83
PASCO SCIENTIFIC 118
PC MAGAZINE 30, 137
PC WORLD 30, 137
PEELEMAN/MC LAUGHLIN ENTERPR 163
PENCILS & PLAY 118
periodicals 122
periodicals for teachers 126
periodicals on science 133
pet care
    NATURE'S TOYLAND 162
petrified wood
    LIZZADRO MUSEUM SHOP 108
PH SCALE 90

PHILOSOPHY OF SCIENCE 138
photography
    NATUREPRINT PAPER 163
physics
    AAPT 84
    AAPT PRODUCTS CATALOG 23
    AMERICAN JOURN OF PHYSICS 127
    AMUSEMENT PARK PHYSICS 8
    ARBOR SCIENTIFIC 111
    DEMONSTRATION EXPERIMENTS 9
    DEMONSTRATION HANDBOOK 9
    FABYAN MUSEUM 67
    FERMILAB 67
    FERMILAB LIBRARY 101
    INTERACTIVE PHYSICS 33
    ISPP 82
    OLYMPICS HANDBOOK 12
    PASCO SCIENTIFIC 118
    PHYSICS BEGINS WITH AN M... 11
    PHYSICS NORTHWEST 83
    PHYSICS OF SPORTS 11
    PHYSICS OF TOYS 163
    PHYSICS TEACHER 130
    PHYSICS TODAY 138
    PHYSICS WEST 83
    SIX FLAGS 60, 75
    TIME MUSEUM 76
    VERNIER SOFTWARE 39
PHYSICS BEGINS WITH AN M... 11
PHYSICS CURRICULUM & INSTRUCTION 168
PHYSICS EXPERIMENTS FOR CHILDREN 11
PHYSICS NORTHWEST 83
PHYSICS OF SPORTS 11
PHYSICS OF TOYS DEMO SET 163
PHYSICS OLYMPICS HANDBOOK 12
PHYSICS TEACHER 130
PHYSICS TODAY 138
PHYSICS WEST 83
PILCHER PARK 74
PIPETTE 90
planetariums
    ADLER PLANETARIUM 65
    ADLER PLANETARIUM SHOP 104
    CERNAN EARTH AND 65
    EXPLORATOY 159
    UNCLE MILTON IND 165
PLASTICS AND THE ENVIRONMENT
    SOURCEBOOK 104
PLASTICS IN OUR WORLD 119
PLAY VISIONS 119, 163
PLAY-TECH, INC. 163

POPULAR SCIENCE 138
POWELL'S BOOKSTORE 22
POWER HOUSE 52, 74
    CATALOG 119
    EDUC PROGRAMS 52
    ENERGY RESOURCE CENTER 100
    GAZETTE 130
    SAFETY 143
    SCI ON SATURDAY 60
precollege
    ACCESS 2000 78
    LEDERMAN SCIENCE 51
    U. S. DEPARTMENT OF ENERGY 87
    VENTURES IN SCIENCE 54
PREHISTORIC LIFE MUSEUM 74
preschool
    BABYBUG 123
    CHICAGO BOTANIC GARDEN 46
    CONSTR PLAYTHINGS 106, 111
    CURIOSITY PLACE 73
    DESIGN SCIENCE TOYS 158
    EARLY CHILDHOOD 9
    EDUCATIONAL TEACHING AIDS 107
    GRAY'S DISTRIBUTING CO. 107
    JOHN G. SHEDD AQUARIUM 49
    KOHL CHILDREN'S MUSEUM 70
    STOREHOUSE OF KNOWLEDGE 109
    WALTER E. HELLER 54
PRESIDENTIAL AWARDS FOR EXCELLENCE IN SCI AND MATHEMATICS 61
PRODIGY 34
professional books 14
PROJECT 2061 16
    SCIENCE FOR ALL AMERICANS 16
PROJECT INFORM 34
PROJECT STAR 119
PROMISING PRACTICES IN MATHEMATICS & SCIENCE EDUCATION 16, 41
PROTRACTOR 90
PUBLICATIONS FOR PARENTS 26
PYRAMID 171
QUAKER OATS COMPANY LIBRARY 101
QUANTUM 125, 130
QUICK SCIENTIFIC TERMINOLOGY 12
QUICK-SOURCE 4
RACE TO SAVE THE PLANET 171
radio
    AMERICAN RADIO RELAY LEAGUE 44
RAIN DOG BOOKS 22
RANGER RICK'S NATURE MAGAZINE 125

RE-ACTIONS 138
READING, THINKING, & CONCEPT DEVELOPMENT 171
references
    BEST SCIENCE AND 2
    BEST SCIENCE BOOKS 2
    CRC HANDBOOK 15
    ENERGY EDUCATION 3
    GENERAL SCIENCE INDEX 122
    GUIDE TO MATH & SCI REFORM 18
    IDEAAS 4
    INFORMATION SOURCES 4
    LIBRARIES UNLIMITED 25
    MC GRAW-HILL ENCYCLOPEDIA 16
    QUICK-SOURCE 4
    SCIENCE SOURCES 1994 7
    SCIENCE TEACHER'S DESK REF 7
    science toys 150
    SCIENCE TRADEBOOKS 7
    STANDARD PERIODICAL DIR 122
    STIS 30
    TRIANGLE COALITION 77
    ULRICH'S 123
research
    CENTER FOR RES 94
RESOURCE 138
resource centers
    FIELD MUSEUM 47, 57
    LAKE COUNTY EDUC CTR 50
    LEDERMAN SCI EDUC CENTER 51
    LEDERMAN SCI EDUC CENTER 51
    NASA TEACHER RESOURCE CTR 52
    NSRC 87
    POWER HOUSE 52
resource reference books 1
RESOURCES FOR SCIENCE 4
RESOURCES FOR TEACHING GEOLOGY 4
RESOURCES FOR THE CLASSROOM 6
RICHARD ADAMIAK 22
RICHTER SCALE 90
RIVER TRAIL NATURE CENTER 74
robots
    TAMIYA AMERICA, INC. 164
ROOSEVELT UNIVERSITY LIBRARY 101
RYDER MATH AND SCIENCE SCHOOL 43
RYERSON CONSERVATION AREA 67

190 Science Fun in Chicagoland

safety 141
    EDMUND SCIENTIFIC 158
    equipment 144
    FLINN CHEMICAL 115
    LAB SAFETY SUPPLY 116
    NATL SAFETY COUNCIL LIBR 100
    SCIENCE FOR THE PEOPLE 139
SAFETY SYSTEM 144
SAFETY TRAINING INSTITUTE 143
SAND RIDGE NATURE CENTER 75
SARGENT-WELCH A VWR COMPANY 119
SAVI/SELPH 119
SCALE OR BATH SCALE 90
SCHOLASTIC SCI PLACE PROGRAM 120
SCHOLASTIC SOFTWARE 39
SCHOOL OF HOLOGRAPHY 48
SCHOOL SCIENCE AND MATHEMATICS 130
SCHOOL SCI AND MATHEMATICS ASSOC 87
schools (K-12) 42
SCHYLLING ASSOCIATES, INC. 164
SCI AND TECHN INFO SYSTEMS 32
SCIENCE 138
SCIENCE & ARTS ACADEMY 43
SCIENCE & TECHNOLOGY IN FACT AND FICTION 6
SCIENCE & TECHNOLOGY INFORMATION SYSTEM 30, 41
SCIENCE & TECHNOLOGY LIBRARIES 92
SCIENCE 2001 TEXT SETS 6
SCIENCE ACTIVITIES 130
SCIENCE AND CHILDREN 130
SCIENCE ASSOCIATION FOR PERSONS WITH DISABILITIES 139
SCIENCE BOOKS & FILMS 138, 168
science clubs
    EDGEWATER, UPTOWN 79
    FUTURE SCIENTISTS 85
    MUSEUM OF SCI & INDUSTRY 83
SCIENCE DISCOVERY CENTER 53
SCIENCE EDUCATION 130
SCIENCE EDUCATION NEWS 130
SCIENCE EXPERIENCES: COOPERATIVE LEARNING 12
SCIENCE EXPLORERS PROGRAM 53
SCIENCE FAIR HANDBOOK 147
SCIENCE FAIR PAPER 147
SCIENCE FAIR PROJECT INDEX 1973-1980 147
SCIENCE FAIR PROJECT INDEX 1981-1984 147
SCIENCE FAIR RESOURCE CATALOG 147

science fairs 146
    IL JUNIOR ACADEMY OF SCI 148
    INTERNATIONAL 61
    SCIENCE FAIRS GRADES 7-12 12
    SCIENCE FAIRS GRADES K-8 12
    teacher's guide to 148
SCIENCE FAIRS AND PROJECTS - GRADES 7-12 12, 149
SCIENCE FAIRS AND PROJECTS - GRADES K-8 12, 149
SCIENCE FARE 16, 151
science fiction
    THE STARS OUR DESTINATION 23
SCIENCE FOR ALL AMERICANS 16
SCIENCE FOR CHILDREN: RESOURCES FOR TEACHERS 6
SCIENCE FOR THE ELEM SCHOOL 17
SCIENCE FOR THE PEOPLE 139
SCIENCE HELPER K-8 CD-ROM 7, 35
SCIENCE ILLUSTRATED 139
SCIENCE IN ELEMENTARY EDUCATION 17
SCIENCE IS... 12
SCIENCE KIT & BOREAL LABORATORIES 120
SCIENCE LINKAGES IN THE COMMUNITY 53
SCIENCE NEWS 139
SCIENCE NEWS: THE WEEKLY MAG 125
SCIENCE OLYMPIAD 87
SCIENCE ON A SHOESTRING 12
SCIENCE POWER 171
SCIENCE PROJECT SURVIVAL GUIDE FOR KIDS AND ADULTS 145
science projects 146
SCIENCE SCOPE 131
SCIENCE SERVICE, INC. 87
SCIENCE SOURCES 1994 7
SCIENCE STARTERS 13
SCIENCE TEACHER 132
SCIENCE TEACHER'S DESK REFERENCE 7
SCI THROUGH CHILDREN'S LITERATURE 18
SCIENCE TRADEBOOKS, TEXTBOOKS AND TECHNOLOGY RESOURCES 7
SCIENCE WEEKLY 125
SCIENCE WORLD 125
SCIENCE-BY-MAIL 53
SCIENCELAND 125
SCIENCELINES 132
SCIENCES, THE 139
SCIENCEWORKS 13
SCIENTIFIC AMERICAN 139
SCIENTIFIC AMERICAN FRONTIERS 171
scientific measurement instruments 88

SCIENTIST, THE 139
scientists
    NEW SCIENTIST 137
    THE NEW EXPLORERS 170
    THE SCIENTIST 139
SCIS curriculum materials
    DELTA EDUCATION 112
SCITECH 53, 75
    CAMP 75
    EDUC PROGRAMS 53
SCITECH BOOK NEWS 139
SCOPE SHOPPE INC. 109
SCOTT RESOURCES 120
SEARLE LIBRARY 101
SEARS ROEBUCK AND CO. 37
SECOND VOYAGE OF THE MIMI 172
SECRET OF LIFE 172
SEISMOMETER 91
SELSI COMPANY, INC. 120, 164
SEMINARY CO-OP BOOKSTORE 22
SHERIDAN MATH AND SCI ACADEMY 43
SHOWBOARD SCI FAIR RESOURCE CAT 147
SIX FLAGS GREAT AMERICA PHYSICS DAYS 60, 75
SKY & TELESCOPE 140
SLING PSYCHROMETER 91
SMALL WORLD TOYS 164
SMITHSONIAN 140
SO YOU WANT TO DO A SCI PROJECT! 147
SOFTWARE PLUS 39
SOFTWARE SOURCE CO., INC. 39
SOMMERVILLE HOUSE 164
sound
    FABYAN MUSEUM 67
space science
    A G INDUSTRIES 155
    AIR & SPACE /SMITHSONIAN 133
    ESTES INDUSTRIES 159
    NASA/NSTA 61
    ODYSSEY 124
    OMNI 137
    SCHYLLING ASSOCIATES 164
    TOYS IN SPACE 14
special needs
    CLEARBROOK CENTER 151
SPECTRUM 132
sports
    PHYSICS OF SPORTS 11
SPRING VALLEY ENVIRON EDUCATION 54
SPRING VALLEY NATURE SANCTUARY 75
SSMART NEWSLETTER 132

ST. JAMES'S HOUSE 109
STANDARD PERIODICAL DIRECTORY 122
standards
    NATL SCI EDUC STANDARDS 41
STANTON HOBBY SHOP INC. 109, 154
STOREHOUSE OF KNOWLEDGE 109, 154
stores - local 104
STRING AND STICKY TAPE EXPMENTS 13
STUDENT SCIENCE TRAINING PROGRAMS 2
summer programs
    ACCESS 2000 44
    CASPAR 45
    CHICAGO CHILDREN'S MUSEUM 46
    SCITECH 53
    SCITECH CAMP 75
SUMMIT LEARNING 120, 164
SUMNER MATH AND SCIENCE ACADEMY 43
SUNBURST 39
SUPERSCIENCE BLUE EDITION 125
SUPERSCIENCE RED EDITION 126
suppliers 103
    NSTA SUPPLIERS 29, 104, 168
supplies - mail order 110
SYLVESTER ELECTRICAL SUPPLY CO. 109
SYNAPSYS SOFTWARE 35
T.H.E. JOURNAL 30
TAMIYA AMERICA, INC. 164
TANDY TECHNOLOGY SCHOLARS 61
TAPESTRY/NSTA 61
TEACH THE MIND, TOUCH THE 13, 64
TEACH YOUR CHILD SCIENCE 13
teacher development
    BIO WEST 78
    CHEM WEST 78
    CHICAGO BOTANIC GARDEN 46
    CHICAGO SYSTEMIC INITIATIVE 46
    EDTALK 15
    ELEMENTARY SCHOOL 15
    GOLDEN APPLE SCI PROGRAM 48
    GUIDE TO MATH & SCI REFORM 18
    IL COMPUTING EDUCATORS 80
    ILLINOIS ASSOC BIOL 80
    INSTITUTE FOR MATHEMATICS AND SCIENCE EDUCATION 49
    ISPP 82
    ISTA 80
    JOURNAL OF SCI TEACH EDUC 131
    LEDERMAN SCI EDUC CENTER 51
    MID-LEVEL NETWORK 82
    NSTA 87
    PHYSICS NORTHWEST 83

192 Science Fun in Chicagoland

PHYSICS WEST 83
READING, THINKING, & 171
SCIENCE DISCOVERY CENTER 53
SCIENCE TEACHER 132
SPECTRUM 132
TEACHERS ACADEMY 54
UNDER THE MICROSCOPE 172
TEACHERS ACADEMY FOR MATHEMATICS AND SCIENCE (TAMS) 54
TEACHING CHEMISTY TO STUDENTS WITH DISABILITIES 18, 143
TEACHING INTEGRATED MATHEMATICS AND SCIENCE 121
TEACHING SCIENCE 132
TEACHING SCI THROUGH DISCOVERY 18
TEACHING SCIENCE WITH TOYS 151
TEAMS 172
technology
    COMPRESSED AIR MAGAZINE 128
    GALVIN LIBRARY 98
    INSTITUTE OF GAS 98
    INTERNATIONAL SOC 85
    INTERNATIONAL TECHN 85
    INTERNL TECHNOLOGY ASSOC 24
    MUDD LIBR FOR SCI AND ENGR 100
    MUSEUM OF SCI & INDUSTRY 72
    SCITECH 75
    TECHNOLOGY TEACHER 132
TECHNOLOGY AND LEARNING MAGAZINE 30, 132
TECHNOLOGY TEACHER 30, 132
TELESCOPE MAKING 140
telescopes
    AMERICAN SCI & SURPLUS 104
    EDMUND SCIENTIFIC 113
    SELSI COMPANY 120
    TOY STATION 109
television programs 169
TEN-MINUTE FIELD TRIPS 13
TERRAPIN SOFTWARE, INC. 39
THE LEARNING TREE 107
THE STARS OUR DESTINATION 23
THERMOGRAPH 91
THERMOMETER 91
thinking
    CAPRICORN TOYS 156
    CRITICAL THINKING PRESS 24
    DAMERT COMPANY 158
    DESIGN SCIENCE TOYS 158
    EXPLORATORIUM STORE 159
    KADON ENTERPRISES 160

    KOLBE CONCEPTS 116, 161
    OWI INCORPORATED 163
    THINKING WORKS 26
THINKING WORKS 26
THOMAS EDISON BOOK OF EXPMENTS 14
THREE RIVERS AMPHIBIAN 120
TIME MUSEUM 76
TIME MUSEUM STORE 23
TIMETABLES OF SCIENCE 18
TIMS 121
    EXTRAPOLATOR 128
    INSTITUTE FOR MATHEMATICS AND SCIENCE EDUCATION 49
TODAY'S CHEMIST 140
TODAY'S SCIENCE ON FILE 140
TOM THUMB HOBBY & CRAFTS 109
TOSHIBA NSTA EXPLORAVISION AWARDS PROGRAM 62
TOTAL SCIENCE SAFETY SYSTEM 144
TOY STATION 109, 154
toys 150
    loan centers 151
    references 150
    sources 155
    toy stores 152
TOYS IN SPACE 14, 151
TOYS R US 109, 154
TRACING THE PATH 168
TRADITION OF SCIENCE 18
TRAILSIDE MUSEUM 76
travel
    JOHN G. SHEDD AQUARIUM 70
TREKS FOR TROOPS 64
TRIANGLE COALITION FOR SCIENCE AND TECHNOLOGY EDUCATION 77
TROST HOBBY SHOP 154
TYCO TOYS, INC. 165
U. S. DEPARTMENT OF ENERGY 87
U. S. ENVIRONMENTAL PROTECTION AGENCY LIBRARY 101
U. S. GOVERNMENT BOOKSTORE 23
ULRICH'S INTERNL PERIODICALS DIR 123
UNCLE FUN 110, 155
UNCLE MILTON INDUSTRIES, INC. 165
UNDER THE MICROSCOPE 172
UNITED STATES GOVERNMENT PRINTING OFFICE 26
units of measurement 88
UNIVERSAL SPECIALTIES CO., INC. 165
UNIVERSITIES RESEARCH ASSOC 101
UNIVERSITY OF CALIFORNIA CENTER 168

UNIVERSITY OF CHICAGO
    LIBRARY 101
    SPECIAL COLLECTIONS 102
UNIVERSITY OF CHICAGO BOOKSTORE 23
UNIVERSITY OF ILLINOIS AT CHICAGO
    SCIENCE LIBRARY 102
UNIVERSITY OF ILLINOIS AT CHICAGO
    BOOKSTORE 23
UNIVERSITY OF ILLINOIS FILM/VIDEO 168
VENTURES IN SCIENCE 54
VERNIER CALIPER 91
VERNIER SOFTWARE 39
video microscopes
    MORITEX 117
video rental
    BLOCKBUSTER VIDEO 166
    COLLEGE VIDEO 166
    UNIVERSITY OF CALIFORNIA 168
    UNIVERSITY OF ILLINOIS 168
video sales
    EDUCATIONAL ACTIVITIES 167
    EDUCATIONAL DESIGN 158
    EUREKA! 24
    HAWKHILL ASSOCIATES 167
    INSIGHT MEDIA 167
    MR. WIZARD INSTITUTE 169
    PHYSICS CURRICULUM 168
    UNIVERSITY OF CALIFORNIA 168
    ZTEK CO. 169
VON STEUBEN METROPOLITAN SCI CTR 44
VOYAGE OF THE MIMI 172
WALTER E. HELLER NATURE CENTER 54, 76
WARD'S NATURAL SCI ESTABLISHMENT 121
water toys
    GALT COMPANY, INC. 160
WEATHERWISE 126
WEIRD SCIENCE KIDS 142
WEST SUBURBAN LAPIDARY CLUB 83
WESTINGHOUSE SCI TALENT SEARCH 62
WHITNEY YOUNG HIGH SCHOOL 44
WIERD SCIENCE KIDS 60
wildlife
    AUDUBON 133
    CRAFT HOUSE CORP. 156
    ILLINOIS AUDUBON SOC 80
    OWL 124
    RANGER RICK'S 125
    THE GARDEN SHOP 106
    WILLOWBROOK 76
WILLOWBROOK WILDLIFE CENTER 76
WIND VANE 91

WISCONSIN FAST PLANTS 121
WIZARD INSTITUTE, MR. 169
Wizard, Mr., science sets
    PLAY-TECH, INC. 163
WIZARD'S, MR., SUPERMARKET SCI 11
women
    ASSOC FOR WOMEN IN SCI 85
WONDERSCIENCE 126
WOOD LIBRARY - MUSEUM
            OF ANESTHESIOLOGY 76
workshops
    BROOKFIELD ZOO 45
    CAMP SAGAWAU 65
    CHICAGO ACADEMY OF SCI 45
    COLUMBIA COLLEGE 47
    FIELD MUSEUM 47
    ILLINOIS DEPT OF ENERGY 48
    INVENTOR'S COUNCIL 49
    SCITECH 53
    TEACHERS ACADEMY 54
YEARBOOK OF SCI AND THE FUTURE 140
young adult books
    SCIENCE IN FACT AND FICTION 6
YOUNG ENTOMOLOGIST'S SOCIETY 121
YOUR BIG BACKYARD 126
ZENITH ELECTRONICS CORPORATION
    TECHN LIBRARY 102
ZOO REVIEW 140
zoology
    BROOKFIELD ZOO 65
    BROOKFIELD ZOO LIBRARY 94
    LINCOLN PARK ZOOL GARDENS 71
    LINCOLN PARK ZOOLOGICAL SOC 82
    N. FAGIN BOOKS 22
    ZOO REVIEW 140
ZTEK CO. 169

About the Author

Thomas W. Sills is professor of physical science at Wright College in Chicago. In high school he received an award at the International Science Fair for his student science project on learning and memory. In the fall of 1967 he taught college physical science for elementary teachers as his first teaching assignment. In 1977 he received his Ph.D. in physics and education at Purdue University.

His diverse professional career in science education includes test development, science toy design, science teacher development, reviewing college physics textbooks, working as faculty coordinator of *The Mechanical Universe* telecourse on Channel 20/Chicago educational television, and acting as science consultant to programs for the gifted.

Dr. Sills is a collector of books and manuscripts on science and technology. Most of all he enjoys going to new places, thinking about new ideas, and doing adventuresome things that are just plain fun.

Corrections, Additions?

If you have found any errors, updated information, or useful additional resources while reading this book, please send them to the author, Thomas W. Sills, c/o Dearborn Resources, P. O. Box 59677, Chicago, IL 60659-0677. Your kind assistance will make the next edition even better. Thanks!

Need another copy?

If you would like a copy of *Science Fun in Chicagoland* and cannot find it at your local bookstore, you can order one from Dearborn Resources. Send a photocopy of this order form to:

Dearborn Resources
P. O. Box 59677
Chicago, IL 60659-0677

FAX 312-262-5731   24 Hours A Day

Quantity

Science Fun in Chicagoland _____   Each $ 14.95 $ _____

Shipping & Handling ($ 3.00 for one copy) $ _____
($ 1.50 per each additional copy)

8.75 % Sales Tax ($ 1.31 per copy) $ _____
(Illinois addresses only)

(30 day money back guarantee!)   TOTAL $ _____

Payment by: ☐ *Check* ☐ *Money Order* ☐ *MasterCard* ☐ *VISA*

Credit Card Number ___ ___ ___ ___   ___ ___ ___ ___   ___ ___ ___ ___   ___ ___ ___ ___

Expiration Date _____   **Signature** _____
(Name on card of buyer printed below)

Sold to: (PLEASE PRINT)   Please allow 3 to 5 weeks for delivery.

Name:_____   Street:_____

City: _____   State: _____   Zip: _____

(Option) Send as a gift to: (PLEASE PRINT)

Name:_____   Street:_____

City: _____   State: _____   Zip: _____